QUALITY OF SERVICE IN A CISCO® NETWORKING ENVIRONMENT

QUALITY OF SERVICE IN A CISCO® NETWORKING ENVIRONMENT

Gilbert Held

4-Degree Consulting
Macon, Georgia, USA

JOHN WILEY & SONS, LTD

Other Wiley Editorial Offices

John Wiley & Sons, Inc., 605 Third Avenue,
New York, NY 10158-0012, USA

WILEY-VCH Verlag GmbH
Pappelallee 3, D-69469 Weinheim, Germany

John Wiley & Sons Australia, Ltd, 33 Park Road, Milton,
Queensland 4064, Australia

John Wiley & Sons (Canada) Ltd, 22 Worcester Road
Rexdale, Ontario, M9W 1L1, Canada

John Wiley & Sons (Asia) Pte Ltd, 2 Clementi Loop #02-01,
Jin Xing Distripark, Singapore 129809

British Library Cataloguing in Publication Data

A catalogue record for this book is available from the British Library

ISBN 0 470 84425 6

Typeset in 10/12pt Bookman by Laserwords Private Limited, Chennai, India.

This book is printed on acid-free paper responsibly manufactured from sustainable forestry, in
which at least two trees are planted for each one used for paper production.

CONTENTS

PREFACE

The aim of this book is to provide Cisco® professionals with a one-stop location to obtain complete and concise information concerning the different aspects associated with obtaining a 'quality of service (QoS)' for applications being transported over local and wide area networks. To accomplish this objective this book explains the concept behind the term 'quality of service'. It then proceeds to walk through the layers of the International Standards Organization Open System Interconnection Reference Model, commencing at the media access control layer to examine how traffic can be prioritized when flowing on a local area network.

Beginning with the flow of traffic on a local area network this book examines techniques to expedite the flow of traffic onto and through the wide area network. Because certain techniques associated with obtaining a QoS capability are better suited for private networks than the public Internet, this book will also examine the suitability of techniques for both private and public networks.

As we will note as we read this book, there are many tools and techniques that can be used in a Cisco networking environment to expedite the flow of information across LANs and WANs. In this book we will first examine each technique in detail to obtain an appreciation of the manner by which it functions. This will be followed by an examination of the use of Cisco equipment to implement the previously described technique. Because this author believes honesty is the best policy, it should be noted in the Preface that the vast product line of Cisco Systems results in differences in the manner by which certain QoS parameters are configured as you move between different switch or router models. This means that while this author will try his best to illustrate how to configure Cisco switches and routers to implement different QoS techniques, his illustrations cannot be expected to cover every equipment configuration. This said, it is important to note that this book covers both concepts and operation. The concept portion of each chapter will introduce you to the rationale behind a particular QoS technique, how it operates and why its use can be expected to enhance one or more parameters associated with obtaining a quality of service. Thus, we can view one portion of a chapter as theory while the second portion of the chapter is focused upon practicality.

If your organization is considering implementing real-time applications, such as Voice over IP (VoIP) and videoconferencing or you need to know how to subdivide traffic into different classes and expedite one class over another, then this book is for you. Thus, regardless of the type of Cisco Systems equipment used by your organization or even if your organization does not use Cisco equipment, you will benefit from the contents of this book.

As a professional author I long ago recognized the value of reader feedback. If you have any comments concerning the contents of this book, feel I omitted coverage of a topic of interest, or feel I dwelt too long on a topic, please put your hands on the keyboard and send me an e-mail. You can reach me at gil_held@yahoo.com.

Gilbert Held
Macon, GA

ACKNOWLEDGEMENTS

Although readers only see the name of the author on the cover of a book, a considerable amount of effort goes into its preparation that resembles a team effort. Thus, this author would be remiss if he did not acknowledge the efforts of many individuals who contributed to the book you are now reading.

Once again I would like to thank Ann-Marie Halligan at John Wiley & Sons Ltd for backing another one of my writing projects. I would also like to thank her assistant, Birgit Gruber, for her efforts expended in obtaining reviews of my book proposal and the issuance of a contract.

Behind the publication of a book is a literal small army of copy-editors, artists, typesetters, jacket designers and the printer. Thus, collectively my hat is off to all of those fine people whose efforts converted my manuscript into this book.

As an old-fashioned author I frequently travel to areas around the globe where hotels typically have electrical sockets that rarely interface with my adapters. I long ago decided that the pen is my preferred writing instrument, however, air turbulence at 30,000 feet as well as my pencil illustrations must be converted into a typed manuscript. Fortunately, I am blessed with the ability to depend upon excellent typists who not only can decipher my handwriting and, in addition, can convert my pencil drawings into professional figures suitable for publication in a book. Thus, once again I am indebted to the efforts of Mrs Linda Hayes and Mrs Susan Corbitt.

Last but certainly not least, a special thanks is due to my wife Beverly. The preparation of a book manuscript is a time-consuming task that requires work during many evenings and weekends. Her kindness, understanding, and support of my writing projects are truly appreciated.

ACKNOWLEDGEMENTS

1

THE MEANING OF QUALITY OF SERVICE

As an introductory chapter our goal is to become acquainted with the contents of this book. In doing so we will first focus our attention upon the title of this chapter, obtaining an appreciation of the meaning of Quality of Service (QoS) in a communications environment. As we describe and discuss the meaning of QoS we will also briefly note some of the techniques that can be employed to favor certain types of frames and packets over other types of frames and packets.

Although many books and trade publications take the liberty of using the terms 'frame' and 'packet' synonymously, we will not. Instead, we will use the term frame to refer to a physical data unit (PDU) flowing at Layer 2 in the International Standards Organization (ISO) Open System Interconnection (OSI) Reference Model. Because frames are primarily delivered via LANs, most of our reference to frames will be with respect to LANs. Similarly, because packets flow at Layer 3 of the ISO OSI Reference Model which primarily effects the interconnection of LANs via WANs, our reference to packets will equate to PDUs flowing on WANs.

In concluding this chapter we will briefly tour the remainder of this book by discussing what lies ahead in subsequent chapters. You can use this information by itself or in conjunction with the Index and Contents to locate information of immediate interest. While this author recommends that this book should be read in its chapter sequence, this author also realizes the time demands of professionals working in the field of communications. Thus, where possible, each of the succeeding chapters was written as independent of one another as possible. This chapter independence makes it possible for readers to turn to a specific chapter of immediate interest without having to read a significant amount of material that may not be applicable to their specific information requirements. That said, grab a Coke, Pepsi or other beverage, your favorite munchies and follow me into the wonderful world of Quality of Service.

1.1 WHY QoS?

If you used one of the emerging chat features added to Microsoft Messenger, Yahoo!Messenger or a similar messaging service, you more than likely noted that the quality of reproduced voice can considerably vary from moment to moment. If you use another voice over IP (VoIP) service, such as Net2Phone, you may encounter similar but not as frequently occurring problems concerning the quality of reproduced voice. For readers not familiar with either type of service let us begin at the beginning and quickly obtain an overview of each.

1.1.1 Using Net2Phone

Figure 1.1 illustrates the dialing keypad of Net2Phone, an alternate communications carrier that is well known for developing a series of VoIP products and whose technology is used in Yahoo!Messenger and other messenger products. Net2Phone provides free PC-to-PC calling, discounted phone-to-phone prepaid calling and free calls within the United States for up to five minutes per call. When you download and install Net2Phone PC-to-Phone software the end result is a dialing pad similar to the one shown in Figure 1.1. Through the use of this dialing pad you can initiate either PC-to-PC or PC-to-Phone calls by selecting one of two buttons shown in the lower left portion of the dialing pad.

Dataflow and control

Although your voice digitized conversation flows in the form of packets from your PC through your Internet Service Provider (ISP) link into the Internet, Net2Phone also operates their own IP network for the transport of digitized voice conversations. This means that the company has a degree of control over their facilities to include deciding when they should upgrade the transmission rate of their leased lines and when they should obtain additional equipment to include routers and voice gateways. The latter provides an interface to the public switched telephone network (PSTN) that enables VoIP calls flowing over

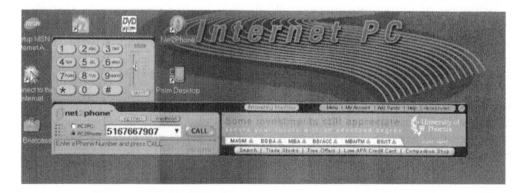

Figure 1.1   The Net2Phone dialpad

the public Internet and Net2Phone's private intranet to be placed onto the PSTN for delivery to Harry Homeowner or Joe Office worker.

1.1.2 Yahoo!Messenger

As mentioned earlier in this section, Net2Phone technology is the power behind several call features added to different messenger products. One such product is Yahoo!Messenger. If you download Yahoo!Messenger and click on its call icon, your screen display will appear similar to the one shown in Figure 1.2. One recent change to Yahoo's calling feature is that the formerly free service now requires the use of a Yahoo phone card from which calls are debited at the rate of 2 cents per minute.

While still a bargain in comparison to the cost of conventional long distance calling, voice digitized calls are first routed via your ISP to Yahoo prior to being placed onto the Net2Phone network for delivery to an applicable gateway.

Dataflow and control

As data flows over the internal Net2Phone network, packets transporting voice are prioritized over packets transporting data. Net2Phone uses a basic method of traffic prioritizing referred to as Type of Service (ToS) coloring in which the bit settings in a field within the IP header are marked to provide a rudimentary signal to network components, such as routers that certain packets should be given high priority over other packets. While this scheme, which is based upon queuing, will be described in Chapter 3 and works reasonably well as packets flow within the Net2Phone intranet, those packets are not differentiated

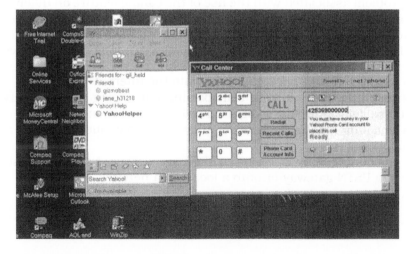

Figure 1.2 Yahoo!Messenger's call facility is similar to several other messenger programs in that it is powered by Net2Phone technology

from other packets as they flow from a user's PC via an ISP access line into and through the Internet to Net2Phone. This means that when you use Yahoo!Messenger where data flows first to a Yahoo location for transfer onto the Net2Phone network, your packets must compete for bandwidth and router resources with other Internet traffic. As you might expect, this results in a series of problems that effect the quality of reproduced voice.

1.1.3 Data transport problems

There are several data transport problems that can occur when packets are not differentiated from one another to allow different levels of service based upon a marking of the packet In this section we will turn our attention to those problems.

Dropped packets

Upon occasion you will note periods of silence during a conversation which results from packets transporting small periods of digitized voice being dropped. While this action may appear shocking to some persons, when routers have more data than they can handle they drop packets. Because real-time communications cannot tolerate the additional delay associated with the retransmission of dropped packets, the alternative is usually to cross one's fingers and hope this does not occur too frequently since the dropping of too many packets could result in an inability to understand the other party.

Distortion of reproduced voice

Another problem associated with the use of the calling feature on different messenger problems is the periodic reconstruction of voice that sounds distorted. That distortion results from the displacement of voice transporting packets from one another due to several factors. Those factors can include the following:

- Delays associated with packets flowing into the Internet
- Delays resulting from packets transporting data being inserted between a series of packets conveying a digitized voice conversation
- Delays encountered by packets as they flow from one router to another through a WAN
- Delays encountered by packets as they are delivered to their destination either via a PSTN gateway or onto a local are a network.

Jitter

The previously mentioned delays result in two distinct types of problems. One problem results from the displacement of packets transporting data from their uniform sequence and is referred to as jitter. If not compensated for by the use of a jitter buffer, at the receiver, reproduced speech may sound like stuttering. Even when a jitter buffer is employed, an excessive amount of displacement of packets from one another in the time domain will result in awkward sounding reproduced voice.

Latency

A second type of problem resulting from delays within local and wide area networks is the one way delay from source to destination that is referred to as latency. As cumulative latency approaches 200 ms or a fifth of a second, a voice conversation will begin to deteriorate, with one party becoming unsure if the other party paused or is waiting for a response. As cumulative latency approaches 250 ms the conversation will begin to require the Citizen Band (CB) use of the keyword 'over' for one speaker to tell the other party they are done rather than have two persons begin speaking at once.

Causes of jitter and latency

Figure 1.3 illustrates the major causes of jitter and latency as a packet flows between one LAN and another via a wide area network. In examining Figure 1.3 let us move from left to right, commencing our examination with the access line into the Internet.

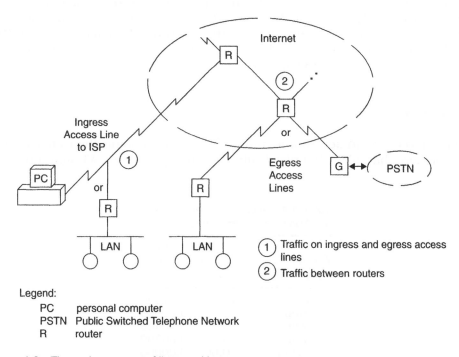

Figure 1.3 The major causes of jitter and latency

The access line

If you are using a PC connected to a LAN to access the Internet frames transporting your digitized voice compete with frames generated by other LAN users for access to the local router and forwarding into the Internet. Although frames leaving your PC that commonly transport 20 ms of digitized voice leave your computer with a uniform gap or time delay between each frame, as they flow onto the LAN they can collide with other frames, enter switches where they overflow a buffer already filled, or even reach the router and be dropped due to excessive activity. Thus, the uniform time delay between frames can be expected to lose their uniformity based upon the occurrence of one or more of the previously mentioned network events.

The operating rate of the access line directly controls latency as well as indirectly contributes to jitter. Concerning latency, as the operating rate of the access line increases, the delay associated with packets reaching the ingress router into the Internet decreases. For example, consider a packet 128 bytes in length. Table 1.1 indicates the latency associated with transporting a 128-byte packet into the Internet at several popular access line data rates.

Note that the operating rate of 1.536 Mbps represents a T1 line. Although the actual operating rate of a T1 line is 1.544 Mbps, 8 Kbps are used for framing and do not transport actual data. Thus, the latency of a 128-byte packet via a T1 access line becomes $128 \times 8/1.536$ Mbps or 0.00066 seconds.

In examining the entries in Table 1.1, note that at a relatively low data rate of 64 Kbps the latency for a 128-byte packet is 16 ms. If the egress access line also operates at 64 Kbps, this means that ingress and egress access lines would result in a total of 32 ms of latency and can represent one of the major contributors to distorted reconstructed voice. Concerning jitter, if the operating rate of the access line is relatively slow, the insertion of a packet transporting data between a sequence of packets transporting digitized voice will have a more pronounced jitter than when the access line operates at a higher rate.

Network flow

As data flows over an access line into an ISPs ingress point, the router at that location more than likely handles traffic from many sources. Without a priority

Table 1.1 Access line latency as a function of line operating rate

Access Line Operating Rate	Latency for 128-byte packet (ms)	(sec)
64 Kbps	16	0.01600
182 Kbps	8	0.00800
256 Kbps	5	0.00400
512 Kbps	3	0.00200
1.536 Kbps	0.66	0.00066

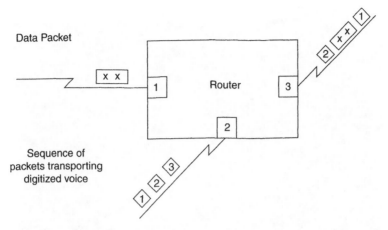

Figure 1.4 The default processing of packets on a first-in, first-out basis can result in packets transporting data being inserted within a sequence of packets transporting digital voice

mechanism, routers service packets on a first-in, first-out (FIFO) basis. This means that a packet transporting data entering the router between two packets transporting digitized voice that are all destined for the same router output port will result in the time displacement of the second packet transporting digitized voice. An example of this situation is illustrated in Figure 1.4.

In examining Figure 1.4 let us assume that the data packet entered the router on port 3 right after the first of the three packet sequence entered the router on port 1. If both packets as well as the following two packets transporting digitized voice are destined to a location that requires packets to egress on port 2, the end result is the insertion of the data packet between the sequence of digitized voice packets. The actual displacement of one packet transporting voice from another depends upon the length of the packet inserted between the two packets transporting digitized voice and the data rate of the communications facility connected to port 2 in this example.

In addition to packets being displaced by time causing jitter, as packets flow through a series of routers they encounter processing delays that add to the latency of the packet. In fact, if you are familiar with the TCP/IP application named traceroute (which is named tracert in a Microsoft environment), you can use that application to view the delays encountered as a packet flows through a series of routers toward its destination. Figure 1.5 illustrates the use of the tracert application to display the router from the router connecting the author's LAN to the Internet and through the Internet to the Yale University Web server.

By default the Microsoft implementation of tracert transmits three IP data-grams with their time-to-live (TTL) field value initially set to one. This action results in the first router in the path to the destination decrementing the TTL value in each datagram by one and noting that the value is now zero, tossing the datagrams into the great bit bucket in the sky. At the same time the router responds to the program origination with an Internet Control Message Protocol (ICMP) time exceeded message that allows the tracert program to compute the round-trip delay. Thus, if you focus your attention upon row 1 in Figure 1.5

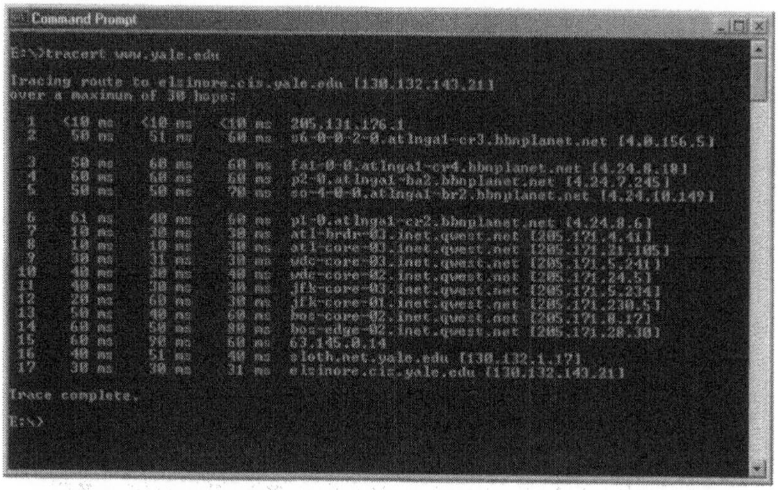

Figure 1.5 The use of the tracert application to display the router from the router connecting LAN to the Internet and through the Internet to the Yale University Web server

you will note three values that represent the round-trip delay to the first router. Next, the tracert program increments the TTL field value by 1 and transmits three additional datagrams as a mechanism to compute the round-trip delay to the second router in the path to the destination.

If you carefully examine the numeric entries in Figure 1.5 you will note a considerable degree of variability in the round-trip delay times both within some rows as well as between rows. This variability results from the fact that currently the vast majority of traffic that flows over the Internet is treated on a first-come, first-serve basis with no quality of service guarantees.

Continuing our examination concerning the causes of jitter and latency shown in Figure 1.3, as packets flow from one router to another they arrive at multiport routers where data streams from numerous sources must be processed and forwarded towards their destination. Thus, additional delays and packet dropping can occur as packets flow through routers towards their ultimate destination. As packets flow between routers, processing delays add to both jitter and latency. Similarly, as packets transporting data are inserted between packets transporting digitized voice, this action also results in an increase in latency and jitter.

Network egress

As packets reach an egress router they are either sent to a gateway for transport onto the PSTN or routed onto an egress access line. As they flow to a gateway and the contents of a snippet of digitized voice is converted back to analog, an additional bit, no pun intended, of delay occurs. Similarly, if packets flow down the access line onto a LAN for delivery to a workstation, they will compete with other LAN traffic that can result in additional delay.

The problem with IP

If we are familiar with the Internet Protocol (IP) we may be aware that IP uses a 'best effort' approach to traffic delivery. This means IP compliant devices by default accept packets on a first-in, first-out basis and do not provide any guarantee about when data will arrive, the order of delivered packets, nor how many packets will be delivered. In addition, because it is possible for IP networks to use different paths for the delivery of packets, they have a degree of unpredictability. In comparison, networks that use frame relay or Asynchronous Transfer Mode (ATM) are based upon the establishment of logically dedicated circuits being established over the physical network infrastructure to interconnect sender and receiver. Although frame relay and ATM can be configured to provide guarantees concerning packet delivery, packet loss, latency, and jitter, the Internet is based upon IP. In addition, because of economics of scale it is usually significantly less expensive to construct a private IP-based network than a frame relay or ATM-based network. Due to the preceding when we discuss Quality of Service we normally are referring to QoS for IP, although many books have been written about QoS in a frame relay and ATM environment. Because most of the QoS techniques developed over the past few years were designed to expedite the transfer of IP, we will primarily focus our attention upon the Internet Protocol in this book. In doing so we will note that QoS techniques primarily fall into three areas. Those areas include reducing the probability that packets will be lost, reducing transfer delay and reducing the delay variations between a sequence of frames or packets.

1.2 DEFINING QoS

Until now we have primarily focused our attention upon data transport problems, noting how jitter, latency and packet loss effect real-time applications but until now deferred a definition for QoS. While QoS is important for making real-time applications work correctly, it is also important for non-real-time applications. After all, would you be willing to sit in front of your computer and observe a file transfer if 99 out of every 100 packets were dropped? Thus, while we think correctly of QoS being critical for real-time applications, we really need some type or level of QoS for all applications.

We can broadly define QoS as 'a service to the intended destination in a timely manner for it to be correctly recognized'. This definition uses the term 'correctly recognized' in place of defining jitter and latency values while the term 'timely manner' is used in place of specifying a packet loss value. Thus, this definition can be used for different applications as long as we recognize the need to know the metrics that make the definition applicable. While the previous definition is rather simplistic, its implementation is not. To obtain an understanding of why QoS can be difficult to put into effect requires us to compare and contrast operations of the public switched telephone network (PSTN) to that of packet networks, so let us do so.

1.2.1 PSTN operations

When we pick up a telephone handset and call a distant party we obtain a quality of service that makes a voice conversation both possible and practical. The practicality of the call results from the basic design of the telephone company network infrastructure. That infrastructure digitizes voice conversations into a 64 Kbps data stream using Pule Code Modulation (PCM) and routes the digitized conversation through a fixed path, established over the network infrastructure. For the entire path, 64 Kbps of bandwidth is allocated on an end-to-end basis to the call. The fixed path is established through the process referred to as circuit switching.

QoS metrics

As voice is digitized at the egress point into the telephone company network, a slight delay of a few milliseconds occurs. As each switch performs a cross-connection operation, permitting digitized voice to flow from a Digital Signal Level Ø(DSØ) contained in one circuit connected to the switch, onto a DSO channel on another circuit connected to the switch, a path is formed through the network and another delay occurs. Although each cross-connection introduces a slight delay to the flow of digitized voice, the switch delay is minimal, typically a fraction of a millisecond or less. Thus, the total end-to-end delay experienced by digitized voice as it flows through a telephone network is minimal.

Another characteristic of the flow of digitized voice through the telephone network infrastructure concerns the variability or latency differences between each digitized voice sample. Although voice digitization and circuit switching processes add latency to each voice sample, that delay is uniform. Thus, we can characterize the telephone network as a low delay, uniform, or near uniform delay transmission system. Those two qualities – low delay and uniform, or near uniform delay – represent two key Quality of Service metrics. The two metrics are important considerations for obtaining the ability to transmit real-time data, such as voice and video. However, the telephone company infrastructure also provides a third key QoS metric, which is equally important. That metric is a uniform, dedicated 64 Kbps bandwidth allocated to each voice conversation. Because the bandwidth is dedicated on an end-to-end basis, you can view it as being similar to providing an expressway that allows a stream of cars to travel from one location to another, while prohibiting other cars destined to other locations to share the highway.

A fourth QoS characteristic provided by the telephone company infrastructure is the fact that digitized voice flows end-to-end, essentially lossless. That is, there is no planned dropping of voice samples during periods of traffic congestion. Instead, when the volume of calls exceeds the capacity of the network, such as on Mother's Day or Christmas Eve, new calls are temporarily blocked and the subscriber encounters a 'fast' busy signal when dialing.

Table 1.2 provides a summary of commonly used QoS metrics and their normal method of representation.

Table 1.2 Common QoS metrics

Metric	Normal Representation
Dedicated Bandwidth	bps, Kbps, or Mbps
Latency (delay)	msec
Variation (jitter)	msec
Data Loss	percentage of frames or packets lost

Although the telephone company network infrastructure provides the QoS necessary to support real-time communications, its design is relatively inefficient. This inefficiency results from the fact that unless humans shout at one another, a conversation is normally half duplex; this results in half of the bandwidth utilization being wasted. Additionally, we periodically pause as we converse. Because 64 Kbps of bandwidth is allocated for the duration of the call, this means that the utilization of bandwidth is far from being optimized.

1.2.2 Packet network operations

In comparison to a circuit-switched network where the use of bandwidth is dedicated to a user, packet networks allow multiple users to share network bandwidth. While this increases the efficiency level of network utilization, it introduces several new problems. To obtain an appreciation of those problems, I will illustrate the operation of a generic packet network; this could be a TCP/IP network such as the Internet, a corporate intranet, or even a Frame-Relay network.

Figure 1.6 shows the flow of data from two different locations over a common backbone packet network infrastructure. In this example, two organizations, labeled 1 and 2, share access via packet network node A to the packet network. Assume that packets destined from organization 1 flow to the network address Z, connected to mode E, while packets from organization 2 flow to location Y, also connected to packet network mode E.

In Figure 1.6, packets could flow over different backbone routes, however, their ingress and egress locations are shown to be in common. Assuming that location 2 is transmitting real-time information to location Y, what happens when a packet from location 1 periodically arrives at node A ahead of the packet from location 2? When this situation occurs, the packet from location 1 delays the processing of the packet arriving from location 2.

Suppose data sources connected to nodes B, C, and D all require access to devices connected to node A. When this situation arises, the device at node A may be literally swamped with packets beyond its processing capability. In this situation, the network device at node A may be forced to drop packets. While applications such as a file-transfer could simply retransmit a dropped packet without the user noticing this situation, if real-time data such as voice or video were being transmitted, too many packet drops would become noticeable; they cannot be compensated for by retransmission that further delays real-time information.

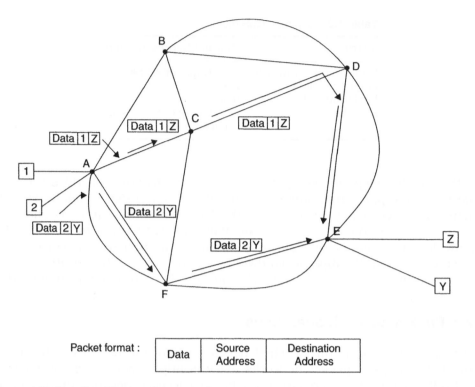

Figure 1.6 Dataflow via a packet network

Consider what happens when the packet from location 2 is serviced at node A. If other packets require routing to note F, packets from location 2 could be further delayed. After packets from location 2 are forwarded onto the circuit between nodes A and F, they will be processed by node F. At this location, packets arriving from nodes C and E could delay the ability of node F to forward packets destined to node E.

Next, packets are forwarded on to node E for delivery to address Y. For the previously described data flow, several variable delays will be introduced, each adversely affecting the flow of packets from location 2 to address Y. Additionally, once a packet reaches node E, it could be delayed by the need to process other packets. Examples include the one arriving from location 1 and destined to address Z.

Another characteristic of a packet network is that when a node becomes overloaded, it will send some packets to the great bit bucket in the sky. This is a normal characteristic of packet networks and, in fact, a frame relay performance metric involves the packet discard rate. Based upon the preceding examination of dataflow in a packet network, note that there is no guarantee that packets will arrive at all or arrive with minimal delay, nor with a set amount of variation between packets. Because this situation makes it difficult if not impossible to transport real-time data over a packet network, various techniques were developed to provide QoS capability to packet networks. Those techniques can be categorized into three general areas. Those areas include expediting traffic at the ingress point into the network, expediting traffic through the network, and expediting delivery of traffic at the destination or egress point in the network. For each area, there are several techniques being supported by different hardware and software vendors to provide QoS capability. Some techniques are standardized, while others will be standardized in the near future.

Implementing QoS

Although standards are important, all vendors do not support all standards. Some standards are currently impractical to implement on a large scale, such as on the Internet. The manner by which an organization connects equipment to the ingress and egress points on a packet network will have a bearing upon whether or not additional QoS tools and techniques are required; they provide an end-to-end transmission capability within certain limits for delay, jitter, and obtainable bandwidth.

As noted by the preceding comparison of the PSTN and packet networks, the manner by which the latter operates makes normal transmission on a 'best effort' basis. When we attempt to implement one or more techniques to provide a QoS capability to a packet network, in effect we are attempting to control the flow of data on that network to the point where bandwidth is reserved similar to the PSTN. This represents a significant challenge since packet networks to include the Internet were designed as a best effort network.

1.2.3 QoS techniques

There are a number of QoS techniques that can be used to expedite certain types of traffic into and through an IP network. One of the most basic techniques is only applicable to a private network. That technique is to adjust available bandwidth and the processing power of routers to minimize packet loss, latency and jitter through the network. Similarly, a technique referred to as the ReSerVation Protocol (RSVP) which allocates bandwidth is only applicable to private networks. Other QoS techniques, such as Differentiated Service (DiffServ), MultiProtocol Label Switching (MPLS) and several types of router queuing are applicable to both public and private IP networks, however, it could be a while until such QoS techniques are available throughout the Internet. In this book we will examine QoS techniques applicable to both public and private IP networks. However, prior to moving on, a few words concerning real-time versus certain non-real-time applications are in order.

Real-time vs. non-real-time applications

While most readers can easily differentiate between real-time applications, such as voice over IP and video conferencing and non-real-time applications, such as e-mail and file transfer, certain applications may appear to represent the

Figure 1.7 Most of the display period you watch on RealPlayer makes it appear that you are watching a real-time presentation

former but in actuality are non-real-time applications. One common example of an application that falls into this category is the use of Real Player from RealNetworks. If you are on a high-speed connection to the Internet it is relatively easy to believe that when you use the program you are receiving a real-time video and/or audio stream. For example, consider Figure 1.7 that illustrates the use of RealPlayer to view a report on death penalty reform from CNN.com.

From observing the RealPlayer screen it appears that the display is occurring at a data rate of 80.5 Kbps and we are 20.5 seconds into a 6 minute and 35.8 second presentation, which is true. However, unless we concentrate our viewing on the lower portion of the RealPlayer screen, it is easy to overlook the fact that data is first buffered prior to the player presenting the data stream.

If you use a relatively slow speed Internet connection you may be able to observe periodic pauses in the display. If you then examine the lower portion of the display you will note the term 'Buffering' followed by the downstream data rate and the time remaining for the buffering operation. An example of a buffering operation is shown in Figure 1.8.

Because a high-speed Internet connection can result in the elimination of most, if not all observable, buffering it may appear that RealPlayer is operating on a real-time basis. However, RealPlayer and similar applications actually are designed to buffer data so they can provide a smooth audio and video presentation. When no or a limited amount of network congestion occurs, RealPlayer will appear to operate as a real-time application, however, it primarily operates as a hybrid between real-time and non-real-time. Thus, RealPlayer can operate in an environment where latency, jitter and packet loss occur. This means that while QoS can be helpful, it is not necessary for the use of RealPlayer and similar applications that use buffering.

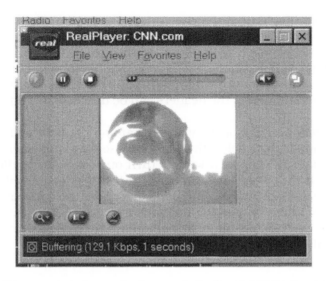

Figure 1.8 RealPlayer and similar applications periodically pause to buffer data

1.3 BOOK PREVIEW

In this concluding section of this chapter we will take a tour of succeeding chapters in this book. You can use the information presented in this chapter by itself of in conjunction with the Index and Contents to locate information of a particular interest. In addition, this brief preview provides a taste of the flow of this book that will indicate the direction this author will take in making readers aware of different QoS techniques and how such techniques can be placed into effect in a Cisco environment.

1.3.1 Working at Layer 2

We begin our examination of QoS in Chapter 2 by commencing our focus upon Layer 2 of the Open System Interconnection Reference Model. Layer 2 of the OSI Reference Model is known as the Data Link Layer and was subdivided by the IEEE into Media Access Control (MAC) and Logical Link Control (LLC) sublayers. As the designated developer of LAN standards, the IEEE also standardized a mechanism to designate the priority of frames on a LAN. That standard, which is referred to as the 802.1p standard, uses a field established by the IEEE 802.1Q standard. Thus, in Chapter 2 we will examine both standards in the first part of the chapter. In the second section of Chapter 2 we will turn our attention to the configuration of Cisco equipment to support the use of priority in a LAN environment.

1.3.2 QoS into the WAN

Because there are several areas within a network where one or more QoS techniques can occur, it is difficult to discuss and describe each as a separate entity by location. In addition, some techniques depend upon other techniques. Thus, as we move off the LAN and into the WAN it becomes possible to implement a variety of QoS techniques. Rather than begin our discussion of QoS into the WAN with a bunch of acronyms relating to different QoS methods, in Chapter 3 we will focus our attention upon the headers of the IP protocol stack and how information in the headers can be used to effect different queuing methods.

In the first section of Chapter 3 we will examine the IPv4 and IPv6 headers, noting, where applicable, fields that can be used for implementing queuing. In the second part of Chapter 3 we will turn our attention to different router queuing methods in a Cisco environment. In this section we will examine the operation of five distinct queuing methods to include the applicable commands required to place different queuing methods into effect. Because queuing is an integral part of any QoS technique this chapter functions as a preview for other chapters that discuss one or more QoS techniques.

1.3.3 DiffServ and MPLS

In Chapter 4 we begin our examination of the flow of data into and through the wide area network. In this chapter we will focus our attention upon two technologies developed to facilitate a QoS capability for IP packets. Those technologies are differentiated services (DiffServ) and MultiProtocol Label Switching (MPLS).

Similar to Chapter 3, our coverage of each technology in Chapter 4 will be in two parts. First, we will focus our attention upon obtaining an appreciation of how the technology operates. Once this is accomplished, we will turn our attention to examining how the technology is implemented in a Cisco environment. Because we are covering two technologies in Chapter 4 this means we will have four sections in this chapter.

1.3.4 The Resource Reservation Protocol

As we move into Chapter 5 we turn our attention to a method that permits bandwidth to be guaranteed for a flow. The method we will discuss is the Resource Reservation Protocol (RSVP).

In the first section of Chapter 5 we will obtain an appreciation for the operation of RSVP. In this section we will note why RSVP is receiver driven, the type of traffic this signaling protocol supports, the key modules used to support the protocol on hosts and routers and RSVP message types and formats. Continuing our prior division of labor, the second section of Chapter 5 is focused upon configuring RSVP in a Cisco router environment. In this section we first discuss the suitability for RSVP based upon the use of different transport facilities. Examining the sequence of RSVP steps required to facilitate the configuration of this signaling protocol will follow this. Once the previously mentioned RSVP-related items have been discussed, we will describe the operation, utilization and format of each RSVP-related command. In concluding this chapter we will examine an example of the configuration of RSVP on a router's serial interface.

1.3.5 QoS enhancement techniques

Until Chapter 6 our primary focus of attention was upon the use of different techniques explicitly developed to provide a QoS capability, such as Differentiated Service, MPLS and RSVP. In Chapter 6 we will turn our attention to a series of eight QoS enhancement techniques that were developed as mechanisms to expedite the flow of traffic into, through and out of a network. The techniques we will cover in Chapter 6 are designed to make transmission more efficient and were not necessarily developed with QoS in mind. However, by considering each of the techniques discussed in Chapter 6 it may be possible for your organization to enhance your overall QoS capability through their utilization.

In Chapter 6 we will examine eight distinct methods that can be used to enhance the flow of data. First, we will examine the use of static routing and how its use eliminates the transmission of routing table updates and their resulting delays to the transfer of other information. This will be followed by an examination of the use of static Address Resolution Protocol (ARP) entries to minimize LAN broadcasts and how the operating rate of access lines can be configured to minimize ingress and egress delays.

Because the use of data compression literally shrinks the amount of data required to be transmitted, its use facilitates the flow of data as well as lowers the level of network utilization. Due to this we will examine the use of RTP and TCP header compression as well as how compression is used in a frame relay environment.

Although this book is oriented towards QoS and not security, we will examine why the elimination of directed broadcasts can help both areas. This will be followed by examining the use of selective acknowledgments and how link fragmentation and interleaving (LFI) can be used to minimize the effect of lengthy packets flowing between packets transporting a real-time application. By examining the eight enhancement techniques discussed in Chapter 6 we may be able to supplement the use of one or more QoS standards to the point where the operation of our applications are better served.

1.3.6 Monitoring your network

In the concluding chapter of this book we turn our attention to a topic that should actually be performed throughout your quest for a QoS capability as well as on a periodic basis after your goal is reached. That topic is network monitoring which is the focus of Chapter 7.

In Chapter 7 we will turn our attention to the literal biblical command used in a Cisco IOS environment for displaying key information about an interface. That command is as you might expect, the show interface command. To ensure all readers have a common level of knowledge, prior to the use of this command we will review the difference between user and privilege modes. Once this is accomplished, we will use the show interface command to examine a router's serial interface and review the metrics provided by the use of that command. Although the show interface command can be used with all interfaces, Chapter 7 restricts its use to the serial interface. The rationale for this restriction results from the fact that the serial interface normally represents the key bottleneck when attempting to obtain a QoS capability. However, readers should also use the show interface command to examine other router interfaces, such as the different flavors of Ethernet, Token-Ring and FDDI as it is possible for highly utilized LANs also to create delays that adversely affect our quest for QoS.

In the second part of Chapter 7 we will turn our attention to the use of IP-related show commands. In the second section of Chapter 7 we will examine how to view the contents of the ARP cache, host statistics, the contents of the route cache, the contents of the routing table and other IP-related information.

2

WORKING AT LAYER 2

Any book covering Quality of Service (QoS) needs to start at the beginning to present a logical view of the technology and terminology associated with the technology. While it is easy to state that we should commence our effort at the beginning, the hard part is attempting to describe where that point is. Some authors like to present material following the historical evolution of a technology. While there can be many positive points associated with this approach, in the area or field of QoS evolution has some serious limitations. Perhaps the most serious limitation results from the fact that QoS is not an all-encompassing technology. Instead, obtaining a QoS capability is typically dependent upon a series of diverse technologies that are brought together to provide a minimum of delay, jitter and dropping of frames and packets on an end-to-end transmission basis. Because QoS technologies were developed for the most part independent of one another, coverage based upon their historical evolution may not be meaningful. Instead, this author will primarily look at the end-to-end flow of data and discuss QoS within those terms, beginning at the LAN.

In this chapter we will focus our attention upon the IEEE 802.1p standard which provides a priority mechanism for traffic at the Open System Interconnection (OSI) Reference Model's Layer 2. That layer, which is referred to as the Data Link Layer, was subdivided by the IEEE into Media Access Control (MAC) and Logical Link Control (LLC) sublayers. The Media Access Control sublayer is located at the lower portion of the split data link layer and is responsible for controlling access to the network. In comparison, the logical link control sublayer is above the MAC layer and is used to provide a connection between network layer protocols and media access control. Because the MAC sublayer provides access to the media we will begin our examination of QoS with an investigation of the operation of the IEEE 802.1p signaling technique at the MAC sublayer in the first section of this chapter. Once we obtain an appreciation of the standard that governs the 802.1p method for prioritizing traffic, we will turn our attention to the use of this signaling technique in a Cisco environment. Thus, the second section of this chapter will be focused on the configuration of Cisco routers and switches to support the IEEE 802.1p signaling technique.

2.1 THE IEEE 802.1p SIGNALING TECHNIQUE

The Institute of Electrical and Electronic Engineers (IEEE) was delegated responsibility by the American National Standards Institute (ANSI) for the development of LAN standards. The actual IEEE committee responsible for LAN standards is the 802 committee, which over the years has developed such well-known specifications as the 802.3 standard for Ethernet and the 802.5 standard for Token Ring LANs. In addition to the 802.3 and 802.5 standards the IEEE has defined a large number of 802 specifications. Thus, a good place to begin when attempting to describe the IEEE 802.1p signaling technique is to review the relationship of the 802.1p standard to other complementary IEEE standards, so let us do so.

2.1.1 Standards relationship

The IEEE 802.1p standard establishes eight levels of priority through the use of a 3-bit extension to a LAN frame. The extension of the LAN frame occurred when the IEEE focused its efforts upon defining a mechanism for the creation of virtual LANs (vLANs). The result of the IEEE vLAN effort was the 802.1Q standard which defines how a tag is inserted into a MAC frame to convey a vLAN identification (vLAN ID). Because it would be more sensible to specify the manner by which LAN frames are modified once instead of twice, the committee developing the 802.1Q standard worked in conjunction with the committee developing the 802.1p specification. As a result of this cooperative effort the vLAN tag specified by the 802.1Q standard has two major parts – a 12-bit vLAN ID field and a 3-bit prioritization field. The 3-bit prioritization field was never actually defined in the vLAN standard. Instead, the 802.1p standard defines the operation and settings of values in the prioritization field.

A third IEEE standard related to 802.1p and 802.1Q standards that warrants mention is the 802.1D standard. Because the 802.1p standard covers traffic class expediting for MAC bridges and the original IEEE 802.1d standard defined bridging operations to include the manner by which the spanning tree algorithm precludes closed logical loops in an Ethernet environment, the 802.1p standard was merged into the 802.1d standard, resulting in the nomenclature of the revision becoming 802.1D. As a popular radio announcer would say, 'Now you know the rest of the story.' Table 2.1 provides a summary of the focus of a portion of the IEEE 802.3 standards supported by many LAN switches and routers.

In examining the entries in Table 2.1 it should be noted that while most LAN switches may support most of the standards listed in Table 2.1, they may not support all standards. For example, a LAN switch that is limited to supporting ports operating at 10 Mbps and 100 Mbps would not support the 802.3z Gigabit Ethernet standard.

Now that we have a general appreciation of the relationship of applicable IEEE standards, let us turn our attention to obtaining an overview of the IEEE 802.1Q standard which allocated a 3-bit field for prioritizing frames.

Table 2.1   IEEE 802.2 standards commonly supported by LAN switches

Standard	Focus
802.1d	Bridge/Spanning Tree
802.1p	Traffic Class Expediting
802.1Q	Virtual LAN
802.1x	Authentication
802.3	Ethernet
802.3μ	Fast Ethernet
802.3x	Auto-negotiation 10/100 Mbps
802.3z	Gigabit Ethernet

2.1.2 The IEEE 802.1Q standard

The IEEE 802.1Q standard specifies the method by which frames are tagged to indicate their membership in a virtual LAN. As a refresher, a vLAN represents a broadcast domain and only a few years ago the amount of hype associated with the evolving vLAN standard appeared to denote something better than the invention of sliced bread. Many of the characteristics of vLANs include the ability to reduce, add, move, and change enhanced security over a large domain via its subdivision into many smaller domains, and reduction of broadcasts since they are limited to each vLAN hold true today. Unfortunately, because the 802.1Q standard is focused on Layer 2 while organizations focus efforts on higher layers of the ISO Reference Model, this has limited the use of vLANs to only a fraction of what was predicted just a few years ago. However, because the frame modification supports prioritization, it is similar to Mark Twain in that its demise is greatly exaggerated.

Overview

The basic definition of a vLAN is 'a collection of network devices grouped together to form a broadcast domain'. The creation of a vLAN occurs within a bridged infrastructure, which means that vLANs architecturally are flat and not hierarchical like a router-based network. As a refresher, a bridge provides connectivity through the forwarding, filtering and flooding of LAN frames. In a vLAN environment a similar 3Fs operation occurs, however, a vLAN switch operates based upon the ports that are within a vLAN configuration. That is, ports within a switch or group of switches are virtually connected as an overlay upon a physical structure of a switch to form a vLAN.

Figure 2.1 illustrates the formation of two vLANs on an 8 port switch. In this example, ports 0, 2, 3, 4, and 5 are assigned to vLAN1 while ports 1,6, and 7 are assigned to vLAN2. Here each port can connect to an individual end station or even a conventional hub.

Advantages of vLANs

The key advantage associated with port-based vLANs is the fact that broadcasts are limited to a vLAN and do not affect devices connected to ports on a different

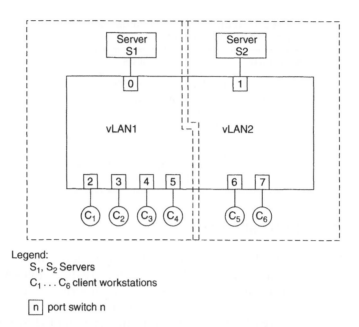

Legend:
    S_1, S_2 Servers
    C_1 ... C_6 client workstations

    ☐ n   port switch n

Figure 2.1   Establishing vLANs based upon the use of switch ports

vLAN. Another advantage is that their use permits managers to group ports and/or users into logically defined communities of interest.

Port-based vLANs are addressed by the IEEE 802.1Q standard. Prior to moving forward it should be noted that vendors also offer network-based vLANs, MAC address-based vLANs, as well as rules-based vLANs that can look deep into the contents of a frame to create vLANs based upon a variety of user-predefined criteria.

Bridges versus switches

If you examine the title of the IEEE 802.1Q standard you will note it references bridged networks but does not contain the term 'switch' in its title. Before you get confused about the lack of reference to switches, it is important to note that the operation of bridges forms the basis for Layer 2 switches and the use of such switches in effect results in the creation of a bridges network. Thus, indirectly the title of the standard references Layer 2 switches.

802.1Q components

The IEEE 802.1Q standard is implemented in LAN switches and not shared media hubs. The standard defines ingress rules that relate to the manner by which frames received at a switch port are classified. Once a frame is classified as belonging to a vLAN to include a default vLAN for all unclassified traffic, the flow of the frame through the switch is based upon forwarding rules. Basically the switch must decide to either filter or forward each frame. A third rule is put into effect when frames are output from a vLAN-compliant LAN switch. At this time the switch must decide if the frame will exit the switch tagged or untagged.

The IEEE 802.1Q specification is defined in approximately 200 technical pages that this author will attempt to summarize in a few pages. The actual flow of frames through a switch depends upon several factors. Those factors include ingress and egress rules that govern the insertion and potential removal of the vLAN tag from a frame, port state information and a filtering database. Port state information can include such configuration information as the presence or absence of a vLAN compatible neighboring switch. This information enables the switch to determine if the vLAN tag should be removed upon the exit of the frame from a switch port as otherwise the extended length of a frame could result in a non-compliant vLAN tag device dropping the extended frame.

A second factor that governs the flow and modification of frames through a vLAN compatible switch is the state of its database. That database associates ports and vLANs and is used by the switch's logic to initiate forwarding decisions.

Figure 2.2 illustrates the major components of the IEEE 802.1Q standard used by switches to add, remove and modify vLAN tags as well as to forward tagged frames through the switch.

Now that we have an appreciation of the major components of the 802.1Q standard and the fact that frames are tagged to become vLAN aware let us turn our attention to the tagging process.

The tagging process

The ability to associate a frame to a vLAN required the expansion of the basic Ethernet frame to accommodate a vLAN identifier. Because the expansion of the basic frame was designed to enable a tag to be placed on each frame, the process is referred to as a tagging process.

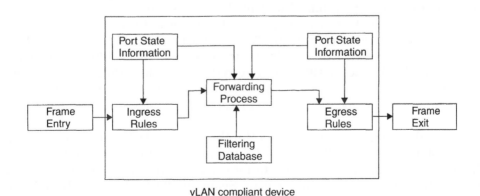

vLAN compliant device

Figure 2.2 Major components of the IEEE 802.1Q standard

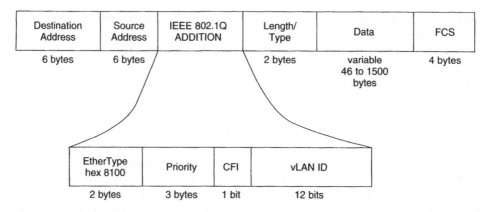

Figure 2.3 Under the IEEE 802.1Q standard Ethernet frames are modified to transport tag control information

Figure 2.3 illustrates the manner by which an Ethernet frame is modified to transport a vLAN tag. Note that tagging results in the insertion of a 4-byte Tag Control Information (TCI) field that includes four subfields.

The TCI field is inserted after the standard 6-byte source field upon an untagged frame entering a vLAN-compliant LAN switch. The first subfield, which is formally referred to as the Tag Protocol Identifier (TPID), defines how the tagged frame is interpreted and is set to a hex value of 8100 in the first two bytes in the tag.

The setting of the two bytes to a value of hex 8100 denotes that the frame includes IEEE 802.1Q/802.1p tagging. Thus, a device reading the first two bytes following the source address field can note from the value of hex 8100 that the next 12-bit positions must be interpreted differently from a named Ethernet frame.

The second major part of the tag is referred to as Tag Control Information (TCI) in IEEE 802.1Q terminology. The TCI contains user priority, a token ring encapsulation flag and a vLAN identifier (VID). The second subfield in the tag and the first in the TCI is a 3-bit priority field whose interpretation is based upon the IEEE 802.1p standard we will shortly discuss.

The third subfield in the tag and the second in the TCI is a 1-bit Canonical Format Identifier (CFI). The 1-bit CFI subfield is set to indicate the situation where an Ethernet frame transports a Token Ring frame across an Ethernet backbone, a situation referred to as Token-Ring encapsulation. The encapsulation process can only occur at points where a frame is either tagged or untagged, locations that correspond to entry and exit ports on a switch.

The fourth subfield in the tag and the third in the TCI is the VID. Through the use of a 12-bit vLAN ID up to 4096 unique virtual LANs can be supported which should be more than sufficient for even the largest type of organization. Because the value of the Frame Check Sequence (FCS) field was based upon the contents of the unmodified frame, once the tagging fields are inserted, the contents of the FCS field must be recomputed.

Now that we have a general appreciation of the manner by which the frame tagging process occurs, let us focus our attention upon the 3-bit priority field and the IEEE 802.1p standard.

2.1.3 The IEEE 802.1p standard

Under the IEEE 802.1Q standard a mechanism was put in place to insert priority information into frames through the use of the 3-bit priority field. The use of a priority mechanism at the MAC layer can be viewed as an initial step in obtaining a QoS capability on an end-to-end basis. The reason for this is due to the fact that the MAC layer does not provide a guaranteed delivery service and it is possible for frames flowing through a switch to be lost. Some of the reasons frames can be lost include a lack of internal device buffering capacity and the length of a frame exceeding the capacity of the destination network. While a priority scheme can be used to overcome the first situation, obviously it has no effect on the second situation.

Need for prioritization

Over the past decade the use of LANs has significantly changed. During the early 1990s the primary use of LANs was to share high cost laser printers and CD jukeboxes, transmit e-mail and access mainframes. Each of these applications were developed for the transport of data and the lack of real-time applications made life relatively easy for network managers and LAN administrators.

Beginning in the second half of the 1990s several applications requiring real-time or near real-time delivery of information began to be developed. By the close of the prior decade applications to include teleconferencing, video conferencing and voice mail retrieval were available for use on LANs. In addition, voice over IP (VoIP) began to gather momentum, resulting in many organizations having to support a variety of real-time and non-real-time traffic.

To provide an applicable level of service to different traffic flows requires the assignment of different priorities to different categories of traffic. This in turn enables LAN switches to prioritize frames based upon a priority assigned to each frame. For example, frames transporting voice over IP would have a high priority to minimize delay and a reconstructed voice sounding like Donald Duck. In comparison, frames transporting e-mail would have a lower priority since the application is relatively immune to an additional second or two of delay. However, to provide frame prioritization frames must first be assigned a priority. To enable hardware and software vendors to develop compatible products a standard level of prioritization became necessary, which is the goal of the 802.1p standard.

Priority settings

Under the 802.1p standard eight levels of priority are defined, ranging from priority 0 through priority 7. Table 2.2 lists the priority levels, with priority 7 being the highest.

Table 2.2   IEEE 802.1p priority values

Bit settings	Decimal value	Description
000	0	normal
001	1	normal
010	2	normal
011	3	normal
100	4	high
101	5	high
110	6	high
111	7	high

In examining the entries in Table 2.2 we can note that this basic classification permits two queues to support priority switching, although additional queues would be welcome under certain conditions. One queue would receive traffic with frames having priority values 000 through 011 while the second queue would receive frames with a priority setting of 100 through 111. Later in this section we will examine how the 802.1p standard can be used to define eight distinct traffic types or classes of service.

Regenerating user priority

As prioritized frames flow into a device their priority setting will be regenerated. Each port includes its own User Priority Regeneration Table that indicates the manner by which the priority value in a frame is regenerated.

Table 2.3 illustrates the default user priority regeneration table. The user priority column indicates the setting of the priority field in the frame upon its entry into a switch. The column labeled 'Default Regenerated User Priority' indicates the new settings for the 3-bit priority field, which by default is the same as the setting in the arriving frame. The third column indicates the range of permitted values. Thus, it is possible to perform a considerable degree of user priority re-mapping if necessary.

Table 2.3   Default user priority regeneration

	Default Regenerated	
User Priority	User Priority	Range
0	0	0–7
1	1	0–7
2	2	0–7
3	3	0–7
4	4	0–7
5	5	0–7
6	6	0–7
7	7	0–7

Traffic classes

As frames are forwarded through an 802.1p compliant network they can be expected to reach a device egress port. That port will have one or more queues into which frames that cannot be directly serviced are temporarily stored. Frames are assigned to each queue based upon their user priority field value. Within each queue unicast frames are given priority over multicast frames.

The actual assignment of frames to a storage queue occurs based upon the use of a traffic class table. Here the traffic class can be considered to represent the number of queues available for each port.

Table 2.4 illustrates the recommended user priority to traffic class mappings defined by the IEEE 802.1p standard. In examining the entries in Table 2.4 some comments are in order to describe its use. If a port only has one queue, then all traffic, regardless of user priority settings, are treated on a first-in, first-out (FIFO) basis. Thus, the mapping of user priority to traffic class 1 results in zeroes regardless of the user priority value.

If the 802.1p compliant port supports two queues, then frames with user priority from 0 through 3 are mapped to traffic class 0 while frames with a user priority value of 4 through 7 are mapped to traffic class 1. Thus, this allows user priorities to be subdivided into two queues if the port is limited to that number of queues. Note that this mapping results in the subdivision of user priorities into two groupings, similar to the priority values shown in Table 2.2. Similarly, when there are eight queues you will note from Table 2.4 that each user priority value is mapped to a separate traffic class.

Traffic types

To facilitate the selection of applicable user priorities the 802.1p standard defines in general terms eight types of network traffic. Table 2.5 lists the types of traffic defined and their relationship to eight levels of user priority.

Table 2.4 802.1p recommended user priority to traffic class mappings

		Number of Available Traffic Classes							
		1	2	3	4	5	6	7	8
U	0	0	0	0	1	1	1	1	2
S									
E	1	0	0	0	0	0	0	0	0
R	2	0	0	0	0	0	0	0	1
P	3	0	0	0	1	1	2	2	3
R									
I	4	0	1	1	2	2	3	3	4
O									
R	5	0	1	1	2	3	4	4	5
I	6	0	1	2	3	4	5	5	6
T									
Y	7	0	1	2	3	4	5	6	7

Table 2.5 Network traffic and user priority

User Priority	Traffic Type
7	a. Network Control/Critical
6	b. Interactive Voice
5	c. Interactive Multimedia
4	d. Controlled Load/Streaming Multimedia
3	e. Excellent Effort/Business Critical
2	f. Best Effort (default) Spare/Standard
1	g. Background

There are two items concerning the entries in Table 2.5 that warrant a few words of elaboration. First, the traffic types listed in Table 2.5 are also referred to as a Class of Service (CoS). In fact, in some literature and frame breakout diagrams the 3-bit priority field in the IEEE 802.1Q tag will be referred to and labeled as the CoS field.

A second item in Table 2.5 that warrants explanation are the letters that prefix the traffic type entries. We will shortly use those letters to indicate how groups of traffic types are grouped based upon the number of queues available in a device.

Turning our attention to the traffic type column in Table 2.5, network control/critical traffic can represent routing table updates and other traffic that have a relatively high requirement to get through to maintain and support the network infrastructure. Frames transporting voice that require less than 10 ms of delay are considered to represent a user priority of 6 while frames transporting video that require a delay under 100 ms represent a user priority of 5. A controlled load or streaming multimedia type of traffic can be considered to represent an important application and is assigned a user priority of 4. In comparison, an Excellent Effort/Business Critical type of traffic is associated with a user priority value of 3. Ordinary LAN traffic can be considered as a Best Effort category and is associated with a user priority of 0, while Background Traffic represents bulk data transfer and is associated with a user priority of 1.

Because ports can support different numbers of queues, as you might expect there exists a mapping between the number of queues supported by port and traffic types. That mapping is indicated in Table 2.6.

2.1.4 Implementing IEEE 802.1p class of service

The ability to implement IEEE 802.1p Class of Service first requires the applicable tagging of frames as they are placed onto the network. While this activity may appear trivial, in actuality it represents a collaborative effort of hardware and software. Concerning hardware, you must use an IEEE 802.1p compliant network interface card (NIC) that is capable of forming extended Ethernet frames that include the four bytes previously described in this section. Concerning software, there are two areas to consider. First, NIC driver software must be obtained that is compatible with the 802.1p standard. Typically the NIC and its drivers are packaged together so this is usually not a problem.

Table 2.6 Traffic type to traffic class mapping

Number of queues	Traffic types
1	a, b, c, d, e, f and g
2	a, b, c, and d
	e, f and g
3	a and b
	c and d
	e and f
4	a and b
	c and d
	e and f
	g
5	a and b
	c
	d
	e and f
	g
6	a and b
	c
	d
	e
	f
	g
7	a
	b
	c
	d
	e
	f
	g

Priority setting

The second type of software required must map applications to a priority value and is a bit more complicated as there are no standards in this area. For example, some application software will work in conjunction with NIC software drivers and pass a priority value to the driver software. Other hardware products, such as the Intel PRO/100+ NIC adapters make it the responsibility of the network manager to develop a mapping policy through proprietary software that works with the manufacturer's NIC. An example of the latter is the Intel Priority Packet software which enables a network administrator to set up priority filters which results in the insertion of a priority value to the applicable 3-bit field in the 802.1Q tag.

In using Intel PRO/100+ NIC adapters the vendor's priority filter is applied to outbound frames. User priority levels 4 through 7 are automatically mapped to the adapter's high priority queue, while levels 0 through 3 are associated with the NIC's standard queue.

Figure 2.4 illustrates the default priority value to queue mapping supported by Intel PRO/100+ NICs. It should be noted that Intel provides a built-in list of predefined filters based upon TCP and UDP port numbers associated

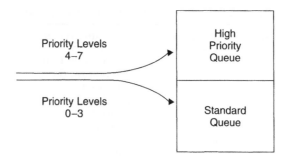

Figure 2.4 By default Intel PRO/100+ adapters will map tagged frames to one of two queues when operating as an 802.10 compliant device

with popular applications, such as Telnet, SNMP and HTTP traffic. Intel also provides a priority filter wizard that facilitates the ability of an administrator to set up custom filters.

Filtering criteria

Intel supports filtering based upon three criteria: node address/Ethernet protocol, TCP/IP and IPX.

The use of node address filtering results in the source or destination Layer 2 MAC address being used for filtering, while the Ethernet protocol filtering uses the two-byte code of the MAC type field. From a practical standpoint the vast majority of filtering can be expected to occur at or above the network layer, resulting in TCP/IP priority filtering being much more commonly used. Since NetWare IPX can be considered to represent a legacy protocol, we will conclude our examination of the Intel PRO/100+ filtering by discussing its TCP/IP filtering capability.

Through TCP/IP filtering the network administrator can prioritize traffic based upon the network or transport layer. At the network layer filtering can occur based upon an IP address and IP subnet mask. At the transport layer the administrator can filter either on all UDP or all TCP traffic or on a specific type of traffic, the latter accomplished by the use of port numbers to specify a specific application.

Microsoft support

The ability to set the priority field value in the 802.1Q tag achieved a high level of support when Microsoft added QoS support to its Windows operating system. At the end of December 1998 Microsoft added what it referred to as 'link layer prioritization' targeted at 802.1p compliance to its Windows 98 and Windows 2000 operating systems. Higher-layer QoS components in the Windows operating system can use directory-based policies, negotiation, or both policies and negotiation with the network to determine applicable priorities. Through the modification of the NDIS packet structure a priority value is placed in a field of the NDIS packet, enabling any network driver to use the priority value to transmit packets with a priority setting.

Windows 2000 admission control

Under Windows 2000 the administrator can centrally designate how, by whom, and when shared network resources are used. This capability is referred to as admission control and works with QoS. Clients and servers running Windows 98, Windows 2000 or Windows XP are automatically configured to use QoS admission control to request bandwidth, however, programs that are not QoS aware will not interact with QoS admission control. This means that if a high level of non-QoS aware applications reside on a network this could severely interfere with the ability to control bandwidth access.

Installation

In a Windows 2000 environment to install QoS Admission Control you would first open the Windows Components wizard. In Components you would click on Network Services and then click on the button labeled Details. Once this is accomplished, you would select the QoS Admission Control Service check box, click on OK, and then click on the Finish button. To perform the previously described operations requires you to be logged in as an Administrator. Figure 2.5 illustrates the previously described process through the selection of Network Services.

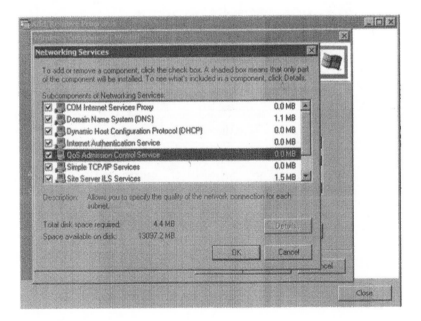

Figure 2.5 Installing QoS admission control services under Windows 2000

Service levels

In forming a QoS admission control policy you need to configure the service level for each data flow. Under Microsoft's QoS admission control three service levels are supported – best effort, controlled load and guaranteed. Best effort represents a setting suitable for elastic traffic that easily adapts to changes in bandwidth availability, such as interactive oriented or bulk data transfer. Controlled load represents a service level that approximates the behavior of best effort in a non-congested condition. This means that a flow receiving controlled load service can be expected to experience little or no delay, congestion or packet loss. The third service level, guaranteed service, provides for a guaranteed maximum delay. Figure 2.6 illustrates the installation of a QoS Packet Scheduler under Windows 2000.

Prioritized traffic flow

Once prioritized frames flow onto the media they need to be recognized and processed according to their tag values. Tag processing can be accomplished in

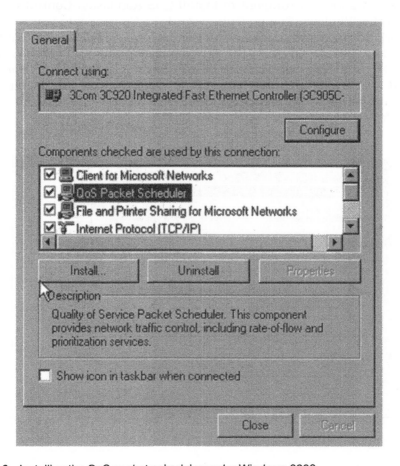

Figure 2.6 Installing the QoS packet scheduler under Windows 2000

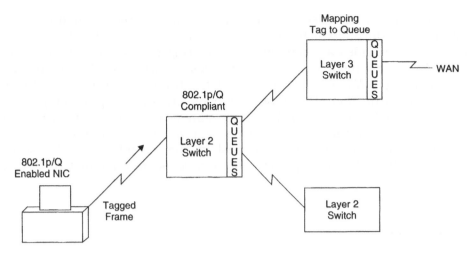

Figure 2.7   Dataflow through Layer 2 and Layer 3 switches

several ways. First, in a strictly LAN environment it is possible to use 802.1p/Q compatible Layer 2 switches. However, if traffic is required to traverse multiple networks, you will more than likely either backend the Layer 2 switch with a router or use a Layer 3 switch to map values to queues for output to one or more WAN links. Figure 2.7 illustrates the relationship between an 802.1p/Q compliant NIC, L2 and L3 switches.

In examining Figure 2.7 note that the Layer 2 switch maps frames by their tag priority field setting into priority queues. However, this action assumes that either the Layer 2 switch or the NIC provided a tagged frame for the Layer 3 switch to operate upon. In actuality there are several variations of tagging or perhaps better put, absence of tagging, that can occur. If the NIC does not tag frames, the Layer 2 switch can be programmed to provide a tag or a pre-defined default tag can be added to frames arriving at a port. If the NIC tags frames, the Layer 2 switch could retag the frame or it can be programmed to use the existing tag to determine what output queue the frame should be placed into. The actual manner by which the Layer 2 switch operates depends upon its support or lack of support of the 802.1Q standard. If it does not support the 802.1Q standard and an untagged frame arrives, the switch will not encounter a problem. However, if a tagged frame arrives that is greater than 1518 bytes in length, the non-compliant 802.1Q switch could consider the frame as an error and send it to the great bit bucket in the sky. Thus, it is wise not to mix 802.1Q compliant and non-compliant equipment in a network.

Now that we have an appreciation for the general implementation of the IEEE 802.1p Class of Service let us focus our attention upon its support in a Cisco environment.

2.2 CONFIGURING CISCO® EQUIPMENT

In this section we will turn our attention to the manner by which Cisco Systems equipment can be configured to support IEEE 802.1p Class of Service

(CoS) values. The 802.1p set of priority values was covered in the first section of this chapter during which we noted its relationship to the IEEE 802.1Q standard that provides a mechanism for creating virtual LANs (vLANs). In a Cisco vLAN environment there are two types of virtual LAN methods used for transmitting vLAN traffic that support the IEEE 802.1p Class of Service method of prioritization. Those two methods are Cisco's proprietary Inter-Switch Link (ISL) and the IEEE 802.1Q standard. In the first part of this section we will turn our attention to configuring Cisco switches for ISL and 802.1Q operation. Once this is accomplished, the second part of this section will examine how the 802.1p Class of Service is implemented in a Cisco environment.

2.2.1 Overview

Although the IEEE 802.1Q standard provides space for the identification of up to 4096 vLANs, Cisco equipment more practically places a limit of support that reflects the fact that keeping track of vLANs requires memory. Depending upon the type of switch used, either 64 or 250 vLANs can be supported on most Catalyst switches. vLANs are identified with a number between 1 and 1001. While most modern Cisco Catalyst switches support both ISL and IEEE 802.1Q trunking methods for transmitting vLAN traffic over 100BASE-T and Gigabit Ethernet ports, not all switches nor switch modules are compliant and readers should note the specifications for the equipment they intend to use.

The command line interface

Cisco equipment supports a command line interface (CLI) that provides you with the ability to configure switches and routers. You can access the CLI either from a console terminal connected to a serial RS-232 console port or via the use of Telnet. Similar to a router, the switch CLI has two modes of operation: normal and privileged. You would enter the normal mode of operation to issue system monitoring commands. In comparison, you would enter the privileged mode to obtain the ability to enter configuration-related commands. Both normal and privileged modes are password protected, enabling two levels of access to switch. Once you enter normal mode you can enter the privileged mode through the use of the enable command followed by an applicable password as shown below:

```
Console>enable
Enter Password:
Console>(enable)
```

For both normal and privileged modes you can view the commands available in a mode by entering 'help' or the question mark ('?'). In addition, appending either 'help' or '?' to a command category or command results in the display of a list of commands in the command category or parameters for a specific command.

Port designation

Most Catalyst switches are modular devices that have ports fabricated on modules. To designate one or more ports requires you to denote both the module and port when working with a modular Catalyst. To specify a port you would enter the module number followed by a forward slash and the port number. If you want to specify multiple ports you can either use a hyphen (-) between port numbers or use multiple entries separated by commas. Table 2.7 indicates four examples of port designations.

vLAN designation

In a Catalyst environment each vLAN is designated by a single number. Similar to the manner by which ports are designated, you can use commas to separate individual vLANs or a hyphen to specify a range of vLANs. For example, 10, 15, 17–20 specifies vLANs 10, 15, 17, 18, 19 and 20.

Core commands

You can configure a Catalyst switch using the set, show and clear commands. The set command is used to initialize or change a switch parameter. You would use the show command to verify the configuration just initialized or changed. The Clear command permits you to overwrite or erase configuration parameters. Note that you can use the set command to set a parameter to a new value.

2.2.2 vLAN port configuration

In a Cisco switch environment a port is associated with a vLAN by assigning a membership mode that determines the type of traffic the port transports and the number of vLANs it can hold membership in. Cisco switches support four types of vLAN membership modes as well as several vLAN membership

Table 2.7 Examples of port designations

Example	Description
3/1, 3/5	Module 3, port 1 and module 3, port 5
3/1–8	Module 3, ports 1 through 8
3/1, 4/6	Module 3, port 1, Module 4, port 6
3/1–3/8	Module 3, ports 1 through 8

combinations. Because only ISL or IEEE 802.1Q types of vLANs support the IEEE 802.1p priority Class of Service we will limit our examination of Cisco vLAN support to those two areas.

For both ISL and 802.1Q vLANs a vLAN Trunk Protocol (VTP) maintains a database that enables the administrator to add, modify and remove vLANs. A trunk represents a connection of two switches and by default is a member of all vLANs in the vLAN database. However, the administrator can alter membership via altering the configuration of an allowed vLAN list or by blocking flooded traffic to vLANs. The latter is accomplished by modifying a pruning-eligible list.

Cisco Catalyst switches construct their address tables by noting the source address of frames received. Similar to a bridge, when a switch receives a frame for a destination address that is not in its address table, it will flood the frame out all ports other than the port the frame was received on. Initially all ports are by default in vLAN1 so frames can be considered to be flooded to all ports of the same virtual LAN except the port the frame was received on. As you configure ports and join them with different vLANs you reduce the effect of flooding since in effect you partition the switch into different broadcast domains.

Switch ports can be assigned to one or more vLANs, with the latter representing a multi-vLAN port.

Figure 2.8 illustrates the use of a switch port to support a multi-vLAN connection from a switch to a router. In this example it was assumed that the switch was partitioned into two vLANs and users on each require the ability to use the services of the router. To assign multiple vLANs to a port to create a multi-vLAN port you would first enter the EXEC mode. Using the CLI you would then perform the sequence of steps listed in Table 2.8. Variables in brackets indicate items you would enter from the console.

In examining the command list in Table 2.8 a few words concerning the assignment of a port to multiple vLANs are in order. When you configure

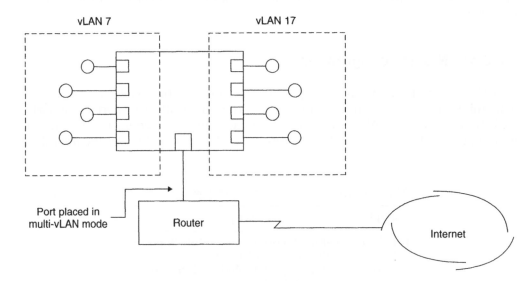

Figure 2.8 Multiple vLANs sharing

Table 2.8 Assigning multiple vLANs to a single port

Command	Description
configure terminal	Enter global configuration mode.
interface<interface>	Enter interface configuration mode and enter the port to be added to the vLAN.
switchport mode multi	Enter vLAN membership mode for multi-vLAN port.
switchport multi vLAN<vlan-list>	Assigns a port to more than one vLAN. Separate nonconsecutive vLAN numbers with a comma, use a hyphen (-) to designate a range of IDs.
end	Return to privileged EXEC mode.

a switch port to support multiple vLANs the vLAN Trunk Protocol (VTP) will transition to transparent mode, which disables VTP. Because an understanding of trunks is essential to appreciating the VTP let us focus our attention on the former prior to investigating the latter.

Trunks

The key to the ability of vLANs to span multiple switches is obtained through the use of one or more trunks. A trunk represents a point-to-point link that connects switches to other switches and routers. As previously mentioned, Cisco switches support two protocols to transmit and receive data via trunks. Those protocols include Cisco's Inter-Switch Link (ISL) and the IEEE 802.1Q. ISL is the default protocol. When using the IEEE 802.1Q protocol there are two limitations that must be considered. First, the native vLAN must be the same at both ends of a trunk link. Second, care must be taken when disabling the spanning tree protocol as the IEEE 802.1Q standard only supports one tree. This means that if you disable STP on the native vLAN of an IEEE 802.1Q trunk without disabling STP on each vLAN in the network you can inadvertently cause loops.

You can also use the CLI to configure a trunk port. In doing so it is important to note that a trunk port cannot be a monitor port or a secure port. When configured as a network port a trunk port functions as the network port for all vLANs.

When configuring a trunk there are several options you must consider, commencing with the protocol (ISL or 802.1Q) to be supported. You also need to consider the mode of the port on the other end of the trunk. To configure an ISL trunk port you would enter the command:

```
set trunk<mode-number/port-number
     (on/desirable/auto/non-negotiate)
```

Once you set the port you can use the show trunk command followed by the module and port number to verify the setting. To configure a port as an IEEE 802.1Q trunk you would add the suffix dot1q to the previously mentioned set trunk command. To verify the trunking configuration you would again use the show trunk command followed by the module number and port number.

When this author was working with certain types of catalyst, switches several 'quirks' were noted concerning the configuration of a port as a trunk. For example, on a Catalyst 5000 the set trunk command always adds all vLANs to the allowed vLAN list for the trunk, even if you specified a vLAN range in the set trunk command. This means that if you only want a subset of all vLANs to be allowed to flow over a trunk you need to use a combination of show, clear and set commands to obtain the desired vLAN allocation to a trunk. That is, you would use the show trunk command to view the vLAN list for the trunks. You would next use the clear trunk command to remove the vLANs you wish to delete being carried via the vLAN. Finally, if required you would use the set trunk command to add specific vLANs to the allowed vLAN list for the trunk. The format of the three commands are indicated below:

```
show trunk<module-number/port-number>
clear trunk<module-number/port-number>
set trunk<module-number/port-number>
```

In concluding our brief discussion of trunks three additional items warrant attention. First, it should be noted that by default all ports are members of vLAN1 regardless of the type of Catalyst switch. Second, many switches include a Graphic User Interface (GUI) which facilitates the configuration of the switch. We can literally kill two birds with one illustration by examining Figure 2.9.

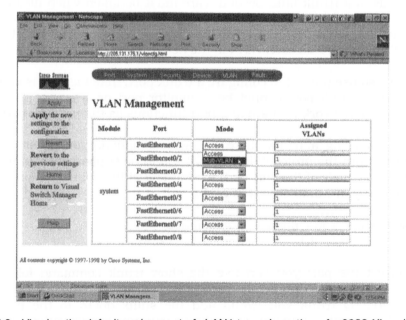

Figure 2.9 Viewing the default assignment of vLAN1 to each portion of a 2908-XL switch

Figure 2.9 illustrates the vLAN Management display of the Cisco Visual Switch Manager for a 2908-XL switch. Note that by default each of the Fast Ethernet ports are assigned to vLAN1.

The third item that warrants attention is a brief description of the spanning tree protocol that governs the active topology of a switch based network.

Spanning tree protocol

If we are familiar with the use of bridges in a legacy Ethernet network we more than likely remember the fact that the spanning tree protocol (STP) is used to define an inverted tree structure that insures there are no closed loops in the active network topology. In an Ethernet environment a closed loop can result in the duplication of frames. Since there is no method to eliminate duplicate frames a closed loop topology not only adds irrelevant network traffic but, in addition, can result in a receiver receiving multiple copies of frames which when processed result in errors within an application. Thus, the use of STP to prevent closed loops both eliminates irrelevant traffic as well as application errors.

In a switch environment the STP is also used to prevent closed loops within an active topology. The left portion of Figure 2.10 illustrates the use of the STP to block the redundant path between switch B and switch C.

In this example the spanning tree protocol represents a logical barrier on a physical topology. Thus, the physical path between switches B and C remains, but the dataflow between the switches is blocked. The right portion of Figure 2.10 illustrates how the spanning tree algorithm will automatically reconfigure the logical spanning tree topology in the event an active link interconnecting two switches becomes inoperative. In this example the link connecting switch A to switch B is considered to have failed, resulting in the link between switch B and switch C moved from a standby into an active state of operation.

A key difference between Cisco's ISL and the IEEE 802.1Q standard concerns the use of the STP. In an IEEE 802.1Q environment a common tree spans all switches regardless of the composition of vLANs. In a Cisco ISL environment you can enable or disable STP on a per-vLAN basis.

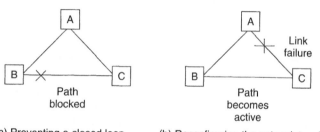

(a) Preventing a closed loop   (b) Reconfiguring the network topology

Figure 2.10   The spanning tree protocol (STP) prevents closed loops in a switch topology

The vLAN trunk protocol

The vLAN trunk protocol provides the mechanism to distribute information about vLANs on a multi-switch basis. To provide multi-switch support VTP maintains vLAN configurations on a network-wide basis as well as functions as a Layer 2 messaging protocol. In doing so VTP operates on a domain or vLAN management domain basis, where a VTP domain represents one or more switches under the same administrative entity.

In a Cisco switch environment a switch will receive an advertisement for a domain via a trunk or via a manual configuration process. If a switch receives a VTP advertisement, it will inherit the domain name in the advertisement, after which it will ignore any advertisements with a different domain name. VTP advertisements include the VTP domain name, configuration revision number, an update identify, and an update timestamp. In addition, the VTP will distribute for each vLAN configured its ID, name, type, state and where applicable, additional information concerning a specific type of vLAN.

VTP configuration

When the VTP is configured for the first time it must be assigned a domain name. That name must also be used as the domain name when configuring other switches in the VTP domain. It is important to note that Cisco publishes guidelines concerning the use of different versions of VTP under different network environments, such as the use of Token Ring networks. As a general rule all switches in a VTP domain must run the same VTP version, however, for other guidelines readers should consult specific Cisco switch manuals.

VTP is configured in a similar manner to the configuration of a multi-vLAN port. That is, you would configure VTP through the switch's command line interface (CLI) by entering commands in the vLAN database command mode.

The ability to change and distribute a new configuration through a network requires a switch to be placed in its VTP server mode. Table 2.9 lists the commands you would enter to configure a Catalyst switch for VTP server mode operation.

In examining the entries in Table 2.9 a few items require elaboration. First, the domain name can be between 1 and 32 characters in length. That name must be the same for all switches operating in VTP server or client mode in a domain. Second, if an optional password is configured, it must be between 8

Table 2.9 Placing a catalyst switch in VTP server mode

Command	Description
vLAN database	Enter the vLAN database mode.
vtp domain<domain-name>	configure a VTP administrative domain name.
vtp password<password>	Set a password for the VTP domain (optional).
vtp server	Configure the switch for VTP server mode.
exit	Return to privileged EXEC mode.

and 64 characters. That password must then be assigned to every switch in the domain for the domain to function properly.

VTP client mode

If a switch is not a VTP server it must be configured as a client. When in the client mode of operation you cannot change the configuration of the switch directly. Instead, the client switch will receive VTP updates from the VTP server in its domain, using the updates to modify its configuration.

Similar to configuring a switch as a server, you need to configure a switch to operate as a client. Table 2.10 lists the commands you need to enter when you are in the privileged EXEC mode to configure a Catalyst switch for VTP client mode operations.

Similar to configuring a VTP server, if you use a password, each switch in the domain must have the same password. The domain name must also be the same for both client and server switches in the domain. Last but not least, the length constraints previously mentioned concerning the domain name and password for the VTP server are also applicable for the VTP client.

Configuring vLANs

Similar to other vLAN-related functions you can configure vLAN information into the VTP database via the use of a command line interface. In this section we will examine the use of the command line interface to add, modify and delete an Ethernet vLAN from a switch that is in VTP server mode. Table 2.11 lists the commands and a description of those commands required for adding, modifying and deleting a vLAN from the VTP database.

For each of the three operations listed in Table 2.11 once you enter the exit command the vLAN database will be updated and propagated throughout the administrative domain. To verify the vLAN configuration you can use the show vLAN name <vlan-name> command and after you add a vLAN, the show vlan<vlan-id> command after you modify a vLAN or the show vlan brief command to verify the removal of a vLAN associated with the port. Table 2.12 lists the commands required to configure a port as a trunk port. The commands listed in Table 2.12 are applicable to both ISL and IEEE 802.1Q trunk ports.

Table 2.10 Configuring a switch for VTP client mode operation

Command	Description
vlan database	Enter vLAN database mode.
vtp client	Configure switch for VTP client mode.
vtp domain<domain name>	Configure a VTP administrative domain name.
vtp password<password>	Optional password
exit	Update vLAN database and return to privileged EXEC mode.

Table 2.11 CLI commands for adding, modifying aid deleting a vLAN from the VTP database

Command	Description
a. Adding a vLAN to the database	
vLAN database	Enter the vLAN configuration mode.
vlan<vlan-id>name<vlan-name>	Add an Ethernet vLAN by assigning a number to it to update the vLAN database.
exit	
b. Modifying a vLAN	
vlan database	Enter the vLAN configuration mode.
vlan<vlan-id>mtu<mtu-size>	Identify the vLAN and change its MTU size.
exit	Update the vLAN database.
c. Delete a vLAN from the database.	
vlan	Enter the vLAN configuration mode.
no vlan<vlan-id>	Remove the vLAN.
exit	Update the vLAN database.

Table 2.12 Configuring a switch port as a trunk port

Command	Description
configure terminal	Enter global configuration mode.
interface<interface_id>	Define port to be configured for trunking.
switchport mode trunk	Configure port as a vLAN trunk.
switchport trunk encapsulation{isl|dot1q}	Configure port to support ISL or 802.1Q.
end	Return to privileged EXEC mode.
copy running-config startup-config	Save the configuration.

Configuring 802.1p class of service

Until now we have primarily focused our efforts in this section towards obtaining an appreciation of the manner by which vLANs are created in a Cisco switched network environment. Using that information as a base we will now examine how we configure 802.1p Class of Service values. However, prior to doing so it should be mentioned that there are differences in the implementation of 802.1p between certain Cisco switches and readers should consult the manual for the specific switch or switches they are using or intend to use to ensure compatibility between switches.

Overview

As previously mentioned, 802.1p is applicable for both ISL and 802.1Q frames. For both protocols frames are classified or tagged for transmission to other devices. As frames flow through a switch-based network, each switch port is limited to a single receive queue buffer as data can only enter a switch serially and the internal processing capability of the switch exceeds the arrival rate of

Table 2.13   Configuring class of service port priorities

Command	Description
`configure terminal`	Enter global configuration mode.
`interface<interface>`	Enter interface to be configured.
`switchport priority default<number>`	Enter priority level 0 to 7.
`end`	Return to privileged EXEC mode.

data. If an untagged frame arrives at a port it will be assigned the value of the port as its port default priority. This value is assigned through the use of either CLI or CMS software. Once assigned, the tagged frame continues to use its Class of Service value as it passes through the ingress port.

Once a frame reaches an egress port it will normally be placed into one of two queues – a normal-priority queue or a high-priority queue. Frames with a priority value of 0 through 3 are placed into the normal-priority queue, while frames with a priority value from 4 through 7 are placed into a high-priority queue. Frames in the normal priority queue are only forwarded after frames in the high-priority queue are services.

Table 2.13 lists the commands you would enter to configure class of service port priorities. These commands assume you are in privileged EXEC mode and should not be confused with a set port level command which is used to determine the order in which ports are given access to the switching bus when two or more ports request access simultaneously.

In examining the use of the switch port command in Table 2.13 note that the CoS value entered determines whether frames are forwarded to the normal or the high-priority queue of the output port. If a priority value level of 0 to 3 is assigned to the port, frames will be forwarded to the normal priority queue of the output port. In comparison, if a priority level from 4 to 7 is assigned, frames will be forwarded to the high-priority queue of the output port.

Switch differences

On some Cisco switches readers need to note that a QoS feature will prioritize traffic based upon the CoS value of a frame and the receive-queue buffer level. For example, on Catalyst 6000 switches each port on the switch has a single receive-queue buffer and four user-defined levels of thresholds for inbound traffic. Depending upon the CoS value and receive-queue buffer occupancy inbound frames may be dropped. Table 2.14 lists the default thresholds for inbound frame dropping based upon CoS settings and receive buffer occupancy.

Although a Catalyst 6000 is similar to other Cisco switches in that outbound traffic goes into one of two queues based upon the CoS value in a frame, the administrator of a Catalyst 6000 can configure how low and high priority transmit queues share the total available transmit queue. In doing so the administrator can alter default drop thresholds under which frames in the low-priority queue with a CoS value of 0 or 1 are dropped when the buffer is 40 percent full and frames with a CoS value of 2 or 3 dropped from the low-priority

Table 2.14 Catalyst 6000 inbound default thresholds

CoS value	Receive-queue buffer occupancy
0 or 1	20 percent or more
2 or 3	40 percent or more
4 or 5	75 percent or more
6 or 7	100 percent full

Table 2.15 Mapping priorities

Layer 2 Class of Service	Layer 3 IP Precedence	DSCP
0	Routine 0	0–7
1	Priority 1	8–15
2	Immediate 2	16–23
3	Flash 3	24–31
4	Flash-override 4	32–39
5	Critical 5	40–47
6	Internet 6	48–55
7	Network 7	56–63

queue when the transmit buffer is 100 percent full. For the high-priority queue the default settings result in frames with a CoS of 4 or 5 being dropped when the transmit buffer is 40 percent full while frames with a CoS of 60 or 7 are dropped when the transmit buffer is 100 percent full. Thus, similar to several notations throughout this chapter, differences in the configuration capability of different switches forces this author to remind readers to reference the specific switch manual for settings applicable for the switch they are using.

Mapping information

In concluding this chapter we will acknowledge that Layer 2 priority is only useful in the local area environment. This means that when frames become packets and flow over the WAN we need a mechanism to map Class of Service to Layer 3. Fortunately, Cisco provides a table referred to as 'Classification' which this author feels is better referred to as mapping that is shown in Table 2.15. Table 2.15 indicates the equivalency between Layer 2 Class of Service, Layer 3 IP precedence and the DiffServe Code Point (DSCP) values, the latter two representing classification methods to accommodate the flow of traffic into and through the wide area network that will be covered in detail in succeeding chapters. For now, it is important to note that Table 2.14 provides a mechanism to map prioritized traffic at Layer 2 to higher transport layers.

3

QoS INTO THE WAN

In the first chapter in this book we noted that the IP protocol stack does not include a QoS mechanism. However, prior to being puzzled by the title of this chapter it should be noted that both the IPv4 and IPv6 headers include mechanisms to differentiate one packet from another. Once packets can be differentiated, it then becomes possible to expedite the flow of one packet over another based upon a marking in a field within the packet. Because the manner used to expedite packets is based upon different types of queuing, the focus of this chapter is upon packet marking and queuing. However, because packet marking and queuing can occur at many areas within a network, we will primarily focus our attention on these two topics in this chapter at the egress location from a local area network into a wide area network. This means that other techniques, such as Differentiated Service (DiffServe) and Multi-Protocol Label Switching (MPLS) that can expedite traffic through a WAN will be covered in later chapters.

In the first part of this chapter we will focus our attention upon the headers in the IP protocol stack as they can be used by themselves or in conjunction with either the contents of a packet or packet length to expedite traffic via different queuing methods. In the second part of this chapter we will turn our attention to Cisco router queuing methods. Although our primary emphasis is upon the LAN to WAN interface in this chapter the queuing methods to be covered are applicable to other locations throughout a network where routers are used.

3.1 THE IP PROTOCOL STACK

If you are familiar with the International Standards Organization (ISO) Open System Interconnection (OSI) Reference Model you more than likely have viewed the seven-layer OSI protocol stack reference model. Because the TCP/IP protocol stack predates the ISO OSI Reference Model it does not follow the seven-layered model. However, the developers of the TCP/IP protocol stack recognized the need for a layered network architecture and designed a five-layer protocol stack, with the application layer representing the top three layers of the OSI Reference Model. Figure 3.1 provides a comparison between the TCP/IP protocol stack and the OSI Reference Model.

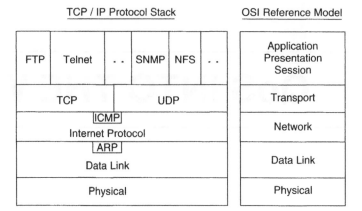

Figure 3.1 Comparing the TCP/IP protocol suite to the OSI reference model

In examining Figure 3.1 note that under the TCP/IP Protocol Stack the application layer represents the upper three layers of the OSI Reference Model. Also note that some protocols in the protocol stack, such as the Internet Control Message Protocol (ICMP) and the Address Resolution Protocol (ARP) represent protocols that operate within Layer 3 (ICMP) and Layer 2 (ARP) to perform specialized functions. Although the TCP/IP protocol stack does not specify a data link nor a physical layer, ARP provides the mechanism to operate over Ethernet, Token-Ring and other data link layers. This in turn enables the TCP/IP protocol stack to operate over different physical layers supported by different data link layers that interface with the Internet Protocol.

3.2 DATA DELIVERY

Without having to probe into the headers of many parts of the TCP/IP protocol stack we can obtain an appreciation of many metrics used for queuing by examining the manner by which datagrams are formed for data delivery. Figure 3.2 illustrates the manner by which a LAN frame containing TCP/IP application data is formed.

The LAN frame header uses a MAC destination address to direct the frame on the local area network. The IP header contains a destination IP address that enables routers to direct the datagram via the wide area network. When the datagram reaches the router connected to the destination LAN, the destination IP address must be equated with a destination MAC address as frames transport packets on the local area network. This action is accomplished by the address resolution protocol (ARP). The router with a datagram to deliver onto the LAN will first check its cache memory to determine if a MAC address was previously learned and associated with the IP destination address in the datagram. If the address was not previously learned, the router will transmit an ARP broadcast frame. That frame contains the destination IP address and is read by each station on the network. The station that recognizes its IP address responds,

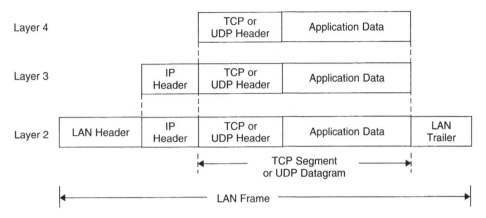

Figure 3.2 LAN delivery of TCP/IP application data

informing the router of its MAC address. This enables the router to place the datagram onto the LAN using the applicable MAC address.

3.3 QUEUING ADDRESSES

Returning our attention to Figure 3.2, important addresses that can be used for queuing include the destination IP address and destination TCP or UDP port numbers. Because the destination TCP and UDP port numbers identify an application, they provide a mechanism to develop a queuing methodology based upon the application being transported.

In addition to the use of the destination IP address and destination TCP or UDP port number, queuing can occur using several additional metrics. Those metrics include the length of a packet as well as certain fields within the IP header. Thus, let us turn our attention to the IPv4 and IPv6 headers.

3.4 THE IPV4 HEADER

The IPv4 header is illustrated in Figure 3.3. This header contains 11 distinct fields plus an option field which, when included may also require a padding field to insure that the option field to include any required padding ends on a 32-byte boundary.

In examining the IPv4 header illustrated in Figure 3.3 we will focus our attention upon the Type of Service (ToS) field as that field is often used for traffic expediting purposes. However, prior to doing so it should be mentioned that it is possible to use other fields in the IPv4 header for traffic expediting purposes. In fact, the 32-bit destination address is commonly used. Less commonly but upon occasion the Identification field which contains a 16-bit number that identifies the following header can also be used to expedite traffic.

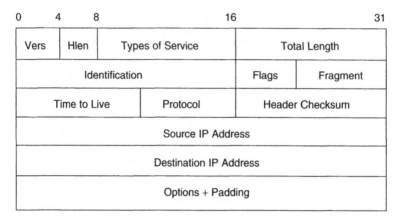

Figure 3.3 The IPv4 header

3.4.1 The type of service field

The Type of Service (ToS) field is one octet or 8-bits in length. The purpose of this field is to denote the importance of the datagram (precedence), delay, throughput, reliability, and cost factors requested by the originator. Although the original intention of including the ToS field in the IP header was to provide a mechanism to differentiate the level of service required for different packets, the use of this field never achieved a significant degree of utilization. One of the key reasons for the lack of use of the ToS field resulted from the fact that its support is optional. Thus, on an Internet basis it is possible for one Internet Service Provider (ISP) to support its use while a second ISP does not. A second reason for its lack of widespread use is the fact that the manner by which routers operate upon values in the ToS field are based upon how the router administrator programmed the device. A third reason for the lack of use of the ToS field is because there is no mechanism in place for ISPs to bill one another for prioritizing one type of traffic over another. As we will note throughout this book, the lack of a billing or charge back mechanism makes it difficult to expedite traffic through an ISP – ISP boundary since the second ISP does not gain anything for expediting the traffic of the first.

Although the use of the ToS field is limited with respect to the Internet, it can and is being used to expedite traffic on private IP networks. In doing so its use is often referred to as 'colored' traffic, where the term color references a particular bit setting in the ToS field. Because it can be and is practically used on the intranet basis let us probe deep and examine its composition.

Composition

Figure 3.4 illustrates the assignment of bit positions within the ToS field. Because the ToS field provides a mechanism to define priorities for the servicing and routing of IP datagrams it can be used to provide a Quality of Service (QoS) capability for IP if all routers along the route support its use in the same

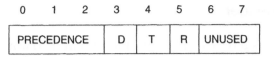

Bits 0–2 : Defines precedence

111 – Network control
110 – Internetwork control
101 – Critic/ECP
100 – Flash override
011 – Flash
110 – Immediate
001 – Priority
000 – Routine

Figure 3.4 The type of service field

manner. Applications can set the appropriate values in the ToS field to indicate the type of routing path they would like. For example, a file transfer would probably request normal delay, high throughput and normal reliability. In comparison, a real time video application would probably select low delay, high throughput and high reliability. While this concept appears to provide a QoS capability, it must be noted that by itself it does not provide a mechanism to reserve bandwidth. Thus, the manner by which a queuing mechanism is implemented as well as how the queuing mechanism operates will have a significant effect upon the ability to provide a QoS capability for colored packets.

Precedence bits

In examining Figure 3.4 note that bits 0 through 2 define the precedence associated with the packet. The top two values are normally network related, however, in a private intranet environment you can define the use of the precedence bits to meet a particular requirement.

Other bits

Bits 3 through 6 govern how packets are treated with respect to delay, throughput, reliability, and cost. A bit setting of 0 indicates no preference or normal treatment, while a bit setting of 1 indicates special consideration, such as low delay, high throughput, high reliability, or low cost. While bit 7 is shown as reserved for future use and set to zero, at one time it should be noted that bit 6 was defined to correspond to a cost metric, with a bit setting of 1 associated with a low cost. Of course, how low is low is quite meaningless and the use of bit 6 for anything other than perhaps a proprietary setting is not commonly used. Thus, in some literature you may see both bits 6 and 7 denoted as reserved for future use.

3.4.2 The IPv6 header

A few years ago it was a reasonable expectation that IPv6 would be implemented on a widescale basis by the turn of the millennium. Well, the turn of the millennium is long over and the use of IPv6 is still in its infancy, so a logical question readers might have is, why should we be concerned with its use? The answer to this question is two-fold. First, the use of IPv6 has not achieved any significant degree due to the architecture of the technology. Instead, the pressing need in the late 1990s for additional IP addresses was satisfied to a large extent through the use of network address translation (NAT) and the Dynamic Host Configuration Protocol (DHCP). Under NAT one Class C address can be used to support up to approximately 65 K private RFC 1918 addresses behind a translation device, significantly extending the use of what are referred to as scarce IPv4 addresses. Under DHCP IPv4 addresses can be dynamically allocated or leased on an as required basis, further extending the use of IPv4 addresses. Between the use of NAT and DHCP the claim of many pundits that all IPv4 available addresses would be in use by the new millennium never materialized and the use of IPv4 continued into the new millennium. Although IPv4 continues to represent the primary Internet Protocol, NAT and DHCP have only extended the day of reckoning and IPv6 will eventually have to be implemented on a large scale.

A second reason for discussing IPv6 is the fact that its header permits both a mechanism to assign a priority to a packet as well as a mechanism to identify a series of packets belonging to the same flow. Both features provide the foundation for providing a QoS capability. This said, let us turn our attention to the IPv6 header.

Header comparison

Figure 3.5 illustrates the IPv6 header. If you compare the IPv6 header to the IPv4 header you will immediately note that other than a common Ver (version)

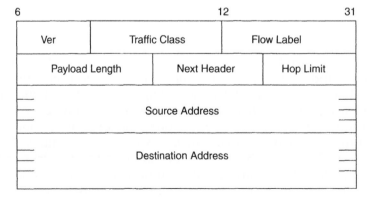

Figure 3.5 The IPv6 header

field they are completely different. Because we are primarily concerned about QoS in this book we will only focus our attention upon certain fields in the IPv6 header in detail, briefly mentioning other fields.

Ver field

The 4-bit Version field contains the binary value of 1010 or decimal 6. This field is the same with respect to position and size as the Ver field in the IPv4 header.

Traffic class field

When the original IPv6 specification was published as RFC 1883 in December 1995 the second field in the header was a 4-bit Priority field. At that time it was envisioned that the 4-bit Priority field would be used to enable a source node, where the term 'node' is used under IPv6 to designate an IPv6 compliant device, to assign one of 16 priorities to a packet. At that time the set of 16 priorities were subdivided into two groups, with values 0 through 7 expected to be used to specify the priority of traffic for which the source is providing traffic control, such as an application using TCP which adjusts its window size during periods of congestion. In comparison, priorities from 8 to 15 were viewed as being used to indicate a range of functions from discarding first (priority 8) to discarding only when absolutely necessary (priority 15). In December 1998 RFC 1883 was made obsolete by RFC 2460 which changed the 4-bit Priority field into an 8-bit Traffic Class field.

Under RFC 2460 it was mentioned that there were a number of experimentations underway in the use of the IPv4 Type of Service and/or Precedence bits to provide for various forms of differentiated service for IP packets. The Traffic Class field in the IPv6 header was noted as being intended to allow similar functionality to be provided in IPv6, however, it was for a future document to detail the syntax and semantics of all or some of the IPv6 Traffic Class bits.

In March 2000 Best Current Practice (BCP) 37 was published entitled *IANA Allocation Guidelines for Values in the Internet Protocol and Related Headers*. In this BCP it was noted that the IPv4 Type of Service field was superseded by the 6-bit Differentiated Services (DiffServ) field plus a 2-bit field that was currently reserved. That BCP also noted that the IPv6 Traffic Class field would also use the 6-bit DiffServ field and a 2-bit reserved field. Thus, because we will focus our attention upon DiffServ in Chapter 4 we will delay an investigation of the manner by which the Traffic Class field bits can be used until that chapter.

Flow label field

The third field in the IPv6 header is the Flow Label field. This field was originally set to 24 bits, but was reduced to 20 bits when the 4-bit Priority field was expanded into an 8-bit Traffic Class field.

The Flow Label field enables a node to label a set of packets that are correlated in some manner, such as having the same source and destination address and TCP or UDP destination port number. From a technical perspective, under IPv6 a flow is uniquely identified by the combination of the source address and a non-zero Flow Label field value. Through the ability to identify a flow it becomes possible for IPv6 compliant routers to handle all packets in a flow in a similar manner.

Payload length field

The fourth field in the IPv6 header is the Payload Length field. This field is 16 bits in length and denotes the length of the data field following the IPv6 header, in octets.

Next header field

The fifth field in the IPv6 header is the Next Header field. This 8-bit field provides IPv6 with the ability to incorporate a string or sequence of headers behind the IPv6 header on an as-required basis. To provide this capability the value in the Next Header field identifies the type of header immediately following the IPv6 header. Similar to IPv4, the two most common kinds of Next Headers will be TCP.6 and UDP.17, however, many other headers are defined whose use are beyond the focus of this book.

Hop limit field

The 8-bit Hop Limit field is similar to the Time-to-Live (TTL) IPv4 field. That is, the value of the Hop Limit field is decremented by one by each node that forwards a packet as a mechanism to prevent packets from endlessly wandering the Internet.

Source and destination address fields

Both the Source and Destination Address fields under IPv6 were expanded from 32 to 128-bit positions. Because each additional bit position doubles the number of potential addresses, the use of IPv6 will provide more than enough distinct addresses for each person on Earth for every gadget imaginable.

Now that we have an appreciation of the IPv4 and IPv6 headers let us turn our attention to router queuing in a Cisco environment.

3.5 ROUTER QUEUING

Any discussion of router queuing methods requires a frame of reference. Because Cisco Systems has approximately 70 percent of the router and LAN switch market, this author decided to describe and discuss router and LAN switch operations throughout this book using examples of Cisco equipment and this chapter is no exception. Thus, in this section as we focus our attention upon router queuing methods we will use Cisco router operations as a frame of reference. In doing so I will describe and discuss queuing algorithms supported by Cisco routers under their Internetwork Operating System (IOS) software. IOS currently supports five different queuing methods – first in, first out (FIFO), priority queuing, custom queuing, weighted fair queuing (WFQ) and class-based weighted fair queuing (CBWFQ), with the latter introduced with the release of IOS 12.1 during 2001. As you might expect, there are certain advantages and disadvantages associated with each queuing method.

3.6 FIRST-IN, FIRST-OUT QUEUING

The first-in, first-out (FIFO) queuing method represents the simplest as well as the default queuing method for interfaces operating above 2 Mbps. The term FIFO is a description of both the manner by which accountants can compute the cost of goods sold, and the flow of data when there is no method available to differentiate traffic.

3.6.1 Operation

Figure 3.6 illustrates an example of FIFO queuing and how this queuing method can result in potential QoS problems that we will shortly discuss. The key advantage of FIFO is the fact that it requires the least amount of router resources. However, the simplistic nature of FIFO queuing is also its key disadvantage. Because packets are output to the interface in the order of their arrival it is not possible to prioritize traffic or to prevent an application or user from monopolizing available bandwidth. To obtain an appreciation for the manner by which FIFO queuing can result in latency problems for latency-intolerant applications, such as Voice over IP (VoIP), let us turn our attention to the use of a voice gateway to digitize voice conversations for their transmission over an IP network.

Returning our attention to Figure 3.6, the voice gateway is shown transmitting relatively short packets on an Ethernet LAN that are indicated by the letter 'v' in a series of square boxes. This is a rather typical example of packet flow during a VoIP conversation, as most applications digitize a sequence of 20 ms of speech into a 64- or 128-byte packet, with the actual packet length dependent upon the application.

3.6.2 FIFO limitations

In the FIFO example shown in Figure 3.6 it is assumed that a station on a Token-Ring network initiates a file transfer operation during a short period of time when a segment of speech is digitized. Assuming that the first packet from

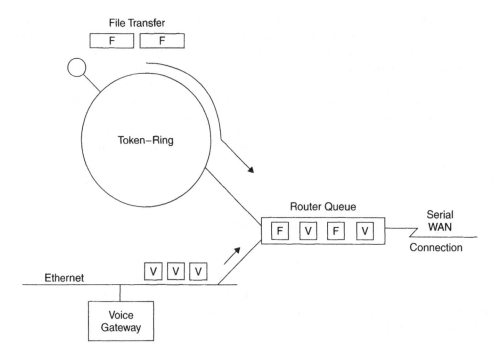

Figure 3.6 Under first-in, first-out (FIFO) queuing frames are extracted from a common queue based upon their order of entry into the queue

the file transfer reaches the router right after the first packet of the digitized voice data stream, a relatively lengthy delay will occur between each segment of digitized voice. Because the reconstruction of digitized voice cannot tolerate any significant delay, such as that associated with the insertion of a lengthy packet between two digitized voice packets in a flow, the data flow in Figure 3.6 also illustrates the necessity of queuing methods that employ multiple queues. By placing traffic into different queues based upon various criteria, and using a priority scheme to extract data from those queues, it becomes possible to differentiate traffic and prioritize different data flows, thereby providing a QoS capability as data flows into the wide area network. Thus, let us continue our investigation of router queuing methods and turn our attention to priority queuing.

3.7 PRIORITY QUEUING

Other than FIFO queuing, priority queuing represents the oldest queuing technique supported by Cisco routers. Under priority queuing traffic can be directed into up to four distinct queues – high, medium, normal, and low. Traffic in the highest priority queue is serviced prior to traffic in the lower priority queues. This means that through priority queuing you can configure a router to place traffic that is relatively intolerant to delay into an appropriate queue that favors its extraction onto the wide area network.

3.7.1 Methods

In a Cisco router environment there are several methods available that can be used to identify traffic to be prioritized as well as to place such traffic into appropriate queues. While it is possible to use an IOS priority-list command by itself to assign traffic to predefined queues, you can also associate an access list with a priority list to take advantage of the considerable flexibility of access lists in filtering data. In this section we will examine the use of the priority list command by itself as well as its use with an access list.

3.7.2 Operation

To illustrate the use of priority queuing and some of its limitations let us first refocus our attention on Figure 3.6. That example illustrated the flow of a file transfer that adversely affected the flow of a stream of digitized voice packets generated by a voice gateway. In that example the two data flows competed for the routers serial port on a first-in, first-out basis. If priority queuing is enabled, it becomes possible to establish multiple queues. Those queues could be configured so that packets conveying the file transfer are directed into the low-priority queue, while digitized voice packets are directed into the high priority queue. Figure 3.7 illustrates an example of this packet direction based upon the use of priority queuing.

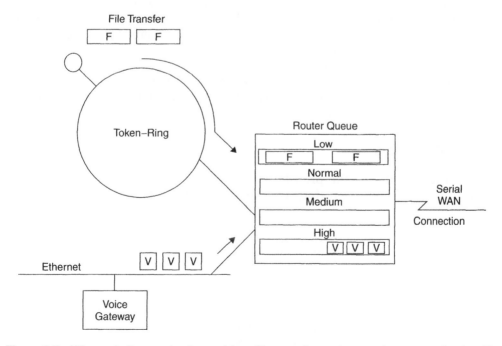

Figure 3.7 When priority queuing is used in a Cisco router environment, you can direct traffic into up to four queues

As previously discussed, voice gateway generated packets are directed into the high-priority queue and file transfer packets flow into the low-priority queue.

The priority-list command

The assignment of traffic to priority queues is accomplished through the use of the priority-list command. That command can be used by itself or in conjunction with an access-list command to direct traffic into applicable queues. When used alone, the format of the priority-list command is as follows:

```
priority-list<list-number>protocol<protocol-name>
[high/medium/normal/low]keyword<keyword-value>
```

Here the <list-number> variable is in the range 1 to 16 and identifies the priority list created. The <protocol-name> variable identifies the protocol type, such as ip, ipx, and similar Layer 3 protocols. Next, select one of the four priority queue levels. This is followed by a keyword that can further define the protocol by specifying a transport protocol carried by the network, such as TCP or UDP. Then, you can specify a <keyword-value> variable that can be used to identify a TCP or UDP port or range of ports.

In addition to TCP and UDP, other keywords supported include fragments, gt, list, and lt. The use of 'fragments' permits the assignment of a priority level to fragmented IP packets. The use of gt (greater than) permits you to prioritize traffic that exceeds a specified byte count. Thus, you would follow 'gt' with a byte count. Similarly, the use of lt (less than) provides you with the ability to prioritize traffic when a packet length is less than the argument value specified after the keyword. Table 3.1 lists the keywords and keyword values you can use with the priority-list command.

Note that the keyword 'list' followed by a list-number provides you with the ability to assign traffic priorities according to a specified access-list, a topic we will look at later in this section.

Utilization example

To illustrate the use of the priority-list command let us assume you want to expedite a VoIP application that uses UDP port 2020. To do so you could use

Table 3.1 Priority-list keywords and values

fragments

```
gt byte-count
list list-number
lt byte-count
tcp port number
udp port number
```

the following priority-list command:

```
priority-list 1 protocol ip high UDP 2020
```

Note in the above example that we use ip as the protocol and specify UDP as the keyword. Because FTP uses TCP ports 20 and 21, the priority-list statements in a Cisco router environment required to configure priority queuing for the two data sources become:

```
priority-list 1 protocol ip low tcp 20
priority-list 1 protocol ip low tcp 21

priority-list 1 protocol ip high udp 2020
```

To effect the filtering of traffic into applicable ports requires the use of a priority-group command. You would use the priority-group command to associate the priority-list to a router's WAN interface. For example, assume the IP address of the serial port connected to the WAN is 205.131.175.1. Then, the applicable set of commands are:

```
Interface serial Ø
Ip address 205.131.175.1 255.255.235.0
priority-group 1
!
priority-list 1 protocol ip low tcp 21
priority-list 1 protocol ip low tcp 20
priority-list 1 protocol ip high udp 2020
```

Although only two types of traffic were assigned to queues, note that any traffic that does not match the priority list entries are, by default, placed in the normal queue.

Now that we have an appreciation of the basic use of the priority-list command, let us focus our attention on two variations of the command. First, we will examine its use with an access list. Once this is accomplished we will examine its use with different router interfaces.

Use with an access list

A second version of the priority-list command can be used to reference an access list. That version of the priority-list command has the following format:

```
priority-list<list-number>protocol<protocol-name>
[high/medium/normal/low] list<list-number>
```

When using the second version of the priority-list command the <list-number> variable references an extended access-list number. To illustrate the use of an access-list in conjunction with a priority-list command the following example indicates how the access-list is used to permit the flow of ftp and udp traffic while the priority-list commands set the traffic into applicable queues.

```
Interface serial Ø
 ipaddress 205.131.175.1 255.255.255.0
 priority-group 1
!
access-list 100 permit tcp any any eq 20
access-list 100 permit tcp any any eq 21
access-list 101 permit udp any any 2020
priority-list 1 protocol ip low tcp 100
priority-list 1 protocol ip low tcp 100
priority-list 1 protocol ip high udp 101
```

3.8 CLASSIFYING TRAFFIC VIA THE ARRIVAL INTERFACE

A third version of the priority-list command can be used to classify traffic based upon the interface it arrives on. The format of this version of the priority-list command is shown below:

priority-list<list-number>**interface**<interface-type><interface-number>[high/medium/normal/low]

In using this version of the priority-list command the <list-number> represents an integer in the range 1 to 16. All statements in one policy would use the same list number. To illustrate the use of this version of the priority-list command let us assume you want to assign traffic entering the router on the ethernet Ø interface to a high priority queue level. This action would be appropriate if the only traffic entering the router on the ethernet Ø interface was from a voice gateway on the Ethernet segment. The statements to expedite traffic from the ethernet Ø port into the high priority queue level are shown below:

```
router(config)# interface e Ø
router(config-if)# priority-list 1 interface serialØ high
```

Queue limits

Prior to moving on, a few words above the maximum number of packets that can be waiting in each queue and how you can adjust default limits are in order. The default queue limits for priority queuing are listed in Table 3.2.

While under ordinary circumstances the packet limits listed in Table 3.2 might appear reasonable, when one or more voice conversations are flowing over an IP network the result is the generation of a relatively large number of short-length packets. Due to the preceding you should consider increasing the default value for the priority queue to which you wish to assign digitized voice packets. To change the queue limits you would use

Table 3.2 Default priority queue packet limits

Priority queue	Default packet limit
High	20
Medium	40
Normal	60
Low	80

the priority list queue-limit command. The format of this command is shown below:

```
priority-list<list-number>queue-limit [high-
limit [medium-limit [normal-limit [low-limit]]]
```

To illustrate an example of the use of the priority-list queue-limit command let us assume you want to set the maximum number of packets in the high priority queue to 100. To do so you would enter the following command:

```
router(config)# priority-list 1 queue-limit 100 40 60 80
```

Limitations

In concluding our discussion of priority queuing it is important to note one of the key disadvantages or limitations of this queuing method. That limitation is the fact that it is possible to starve certain applications from access to an interface, such as the WAN in Figure 3.6. The reason for this is that any time there are packets in the high priority queue, they will be extracted first. Therefore, if the voice gateway is heavily used and generates a lot of traffic on a sustained basis to the router, it is possible to exclude the servicing of traffic entering other queues. In this situation the other queues could fill to capacity, resulting in packets to those queues being dropped. This action could result in the retransmission of packets until a threshold is reached that would terminate the application. Perhaps because of the limitation of priority queuing Cisco Systems added support for three additional queuing methods to its router platforms. One of those methods that should logically be considered after priority queuing is custom queuing. Thus, let us turn our attention to this topic.

3.8.1 Custom queuing

Custom queuing considerably expands queuing on an interface as it subdivides the bandwidth on an interface into 17 queues. Queue Ø represents the system queue and is always serviced first. Into this queue system related packets, such as keepalives and other critical interface traffic are directed. The remaining traffic can be assigned to queues 1 through 16, which are serviced in round-robin fashion commending with queue 1.

Operation

Custom queuing operates based upon specifying a byte count in a series of queue-list commands. By defining the number of bytes to be extracted from a queue before having the router process the next queue, you obtain the ability to indirectly allocate bandwidth. In a Cisco IOS command environment you would use the queue-list command to classify traffic as well as assign the byte counts to specify the maximum quantity of data to be extracted from each queue prior to the next queue being serviced. Because a router cannot transmit a portion of a packet, it uses a byte counter for each queue to keep track of any excess transmission required to complete the packet. When that particular queue is serviced again, the byte count that exceeded the threshold is subtracted from it, in effect penalizing the queue by the number of bytes that it went over its quota. While the adjustment occurs on the next round-robin service of the queue, it provides an additional level of fairness to the custom queuing mechanism.

Queue-list commands

The commands for custom queuing are very similar to that for priority queuing. You can use the queue-list queue command followed by a byte count to specify the number of bytes to be extracted from a queue during a round-robin cycle. Other commands include a queue-list queue limit command that is used to designate the queue length limit for a queue, a queue-list protocol command which enables queuing based upon protocol type, a queue-list interface command that establishes queuing priorities on packets entering an interface, and a queue-list default command that is used to set a priority queue for those packets that do not match any other rule in the queue list.

By obtaining the ability to define the number of bytes to be extracted from a queue prior to having the router process the next queue, you obtain the ability to indirectly allocate bandwidth. For example, let us assume you want to allocate 60 percent of the bandwidth of the WAN connection previously shown in Figure 3.6 to the voice gateway application, 30 percent to file transfers, and the remaining 10 percent to all other traffic. In a Cisco IOS command environment you would use the queue-list command to classify traffic as well as assign the byte counts to specify the maximum quantity of data to be extracted from each queue prior to the next queue's being serviced. There are two versions of the queue-list command you could consider for the previously mentioned situation. You could allocate bandwidth based upon the protocol type or the byte size. We will first examine the use of these two versions of the queue-list command and then examine the other versions of the command.

Custom queuing by protocol

The format of the queue-list command to implement custom queuing by protocol is shown below:

```
queue-list<list-number>protocol<protocol-
name>queue-number><keyword><keyword-value>
```

The first four entries in the queue-list protocol command function in the same manner as the first four entries in the priority-list protocol command. That is, after the keyword queue-list you would enter a <list-number> value between 1 and 16 to identify the list. That entry is followed by the keyword 'protocol', which is followed by an argument that specifies the protocol type, as ip or ipx. The fifth entry, <queue-number>, is an integer between 1 and 16 and represents the number of the queue. The <keyword> and <keyword-value> function in a similar manner as their counterparts in the priority-list command. For the queue-list protocol command possible keywords include lt (less than), gt (greater than), list, tcp and udp.

Custom queuing by byte count

A second version of the queue-list command that can be used is the queue-list queue byte-count command. The format of this version of the queue-list command is:

```
queue-list<list-number>queue<queue-number>byte-
count<byte-count-number>
```

In this version of the queue-list command, the <list-number> and <queue-number> variables continue as identifiers for the number of the queue list and the number of the queue, respectively. The important change is the addition of the keyword 'byte-count', which is followed by the <byte-count-number> variable, the latter specifying the normal lower boundary concerning the number of bytes that can be extracted from a queue during an extraction cycle. The queue extraction process proceeds in a round-robin order, and up to 16 distinct queues can be specified under custom queuing.

The custom-queue-list command

Once you use the queue-list protocol or queue-list-byte-count commands, you would also need to use the custom-queue-list command to associate a custom queue list to an interface. The format of the custom-queue-list command is:

```
custom-queue-list <list>
```

where list represents a number from 1 to 16 for the custom-queue list.
    Returning to our example previously shown in Figure 3.6, our custom-queuing entries, assuming the use of the serial Ø router port for the WAN connection whose IP address is 205.131.175.1, would be:

```
Interface serial Ø
ip address 205.131.175.1 255.255.255.0
```

```
  custom-queue-list 1
  !
queue-list 1 protocol ip 1 udp 2020
queue-list 1 protocol ip 2 tcp 20
queue-list 1 protocol ip 2 tcp 21
queue-list 1 default 3
queue-list 1 queue 1 byte-count 3000
queue-list 1 queue 2 byte-count 1500
queue-list 1 queue 3 byte-count 500
```

In this example, custom queuing was configured to accept 3000 bytes from the UDP queue, 1500 bytes from the ftp queue, and 500 bytes from the default queue. Note that this allocates the percentage of available bandwidth as 60, 30 and 10 percent to queues 1,2, and 3, respectively. Also note that during the round-robin queue extraction process, if there are fewer than the defined number of bytes in a queue, the other queues can use the extra bandwidth. That use will continue until the number of bytes in the queue equals or exceeds the specified byte-count for the queue.

Other queue-list commands

In our prior discussion of custom queuing we briefly noted there are a series of related queue-list commands. In this section we will briefly look at additional queue-list commands.

Queue-list default

The queue-list default command is used to assign a priority for packets that do not match any other rule in the queue list. The format of this command is:

queue-list<list-number>default<queue-number>

Here the <list number> identifies the queue list while the <queue-number> represents the number of the queue, with both numbers being in the integer range of 1 to 16.

Queue-list interface

The queue-list interface command is used to establish queuing priorities on packets entering a router interface. The format of this command is:

queue-list<list-number>**interface**<interface-type>
<interface-number><queue-number>

The following example illustrates the use of the queue-list interface command to assign a priority for packets entering the router's eØ interface to queue number 5. In this example the assignment occurs via queue-list 7:

```
queue-list 7 interface e0 5
```

Queue-list queue limit

The queue-list queue limit command provides a mechanism for assigning a queue length limit to a queue in terms of packets that can be enqueued. The format of this command is shown below:

queue-list<list-number>**queue**<queue-number>**limit**<limit-number>

In this command the <limit-number> variable represents the maximum number of packets that can be enqueued at any time. The range is 0 (queue is unlimited) to 32767 queue entries, with the default value 20 entries. The following example illustrates resetting the queue length for queue 1 to 50:

```
queue-list 2 queue 1 limit 50
```

Similar to our discussion concerning priority queuing, if you are using a VoIP application that generates a large sequence of short-length packets to transport digitized voice, you should consider adjusting the default queue length. Otherwise, it is quite probable under heavy loading for the queue assigned for the VoIP application to rapidly fill and result in subsequent packets being lost during the round-robin process.

Limitations

In addition to the number of packets you need to carefully consider the length of each frame under custom queuing to obtain the desired allocation of bandwidth. This is due to the manner by which TCP operates, as well as how queuing extraction works. Concerning the former, if you are using a TCP application where the window size for a protocol is set to 1, then that protocol will not transmit another frame into the queue until the transmitting station receives an acknowledgement. This means that if your byte count for the queue was set to 1500 and the frame size is 512 bytes, then only about one-third of the expected bandwidth will be obtained, since 512 bytes will be extracted from the queue. As earlier noted, entire frames are extracted, regardless of the byte-count value for a queue. This means that if you set the byte count to 512, but the frame size of the protocol assigned to the queue is 1024 bytes, then each time the queue is serviced, 1024 bytes will be extracted. This would therefore double the bandwidth used by this protocol each time there was a frame in an applicable queue when the round-robin process selected the queue. Therefore, you must carefully consider each protocol's frame size when determining the byte count to be assigned to a queue. Although custom queuing can prevent the potential starvation of lower priority queues, the actual allocation of bandwidth may not be capable of matching your desired metric.

Another limitation of custom queuing concerns the fact that this queuing method requires processing byte counts for up to 16 queues. In doing so custom queuing requires more processing power than priority queuing. Perhaps because of the first problem in which the allocation of bandwidth may not meet your desired metric, Cisco provided two additional methods that can be used to achieve different levels of fairness in allocating bandwidth. Those methods are Weighted Fair Queuing (WFQ) and Class-Based Weighted Fair Queuing (CBWFQ).

3.8.2 Weighted fair queuing

Weighted Fair Queuing (WFQ) represents an addition to IOS that appeared in IOS Version 12.0 on high-end interfaces and router models. In comparison to priority and custom queuing, WFQ represents an automated method to obtain a level of fairness in allocating bandwidth.

Overview

Under Weighted Fair Queuing, all traffic is monitored and conversations are subdivided into two categories – those requiring large amounts of bandwidth, and those requiring relatively small amounts of bandwidth. This subdivision results in packets queued by flow, with a flow based upon packets having a common setting, such as the setting of the Type of Service (ToS) bits in the IP address, and the source and destination TCP or UDP ports. The goal of WFQ is to ensure that low-bandwidth conversations receive preferential treatment in gaining access to an interface, while permitting the large bandwidth conversations to use the remaining bandwidth in proportion to their weights.

Under WFQ, response times for interactive query-response and the egress of small digitized voice packets can be improved when they are sharing access to a WAN with such high-bandwidth applications as HTTP and ftp. For example, without WFQ, a query to a corporate Web server residing on a LAN behind a router could result in a large sequence of lengthy HTTP packets flowing into the WAN link that precede a short digitized voice packet in the interfaces FIFO queue. This would cause the digitized voice packet to wait for placement onto the WAN that could induce an unacceptable amount of delay that would adversely affect the ability to reconstruct a portion of the conversation being digitized. When WFQ is enabled, the digitized voice packet is automatically identified and scheduled for transmission between HTTP frames, which would considerably reduce its egress time onto the WAN.

Implementation

Another advantage of WFQ is that it can be implemented without any options or configuration commands. That is, it is simply enabled by the use of the following interface command:

```
fair-queue
```

Thus, to implement WFQ on serial port Ø you would enter:

```
interface serial Ø
fair-queue
```

Under WFQ, traffic from high-priority queues is always forwarded when there is an absence of low-priority traffic. Because WFQ makes efficient use of available bandwidth for high-priority traffic and is enabled with a minimal configuration effort, it is the default queuing mode on most serial interfaces configured to operate at or below the E1 data rate of 2.048 Mbps.

Options

Although WFQ does not require any configuration commands, the fair-queue command has three options that are noted in the following command format:

```
fair-queue [congestive-discard-threshold[dynamic-
queues[reservable-queues]]]
```

The <congestive-discard-threshold> represents the number of messages allowed in each queue. The default value of this option is 64, with its allowable range 1 to 4096.

The <dynamic-queues> options represents the number of dynamic queues that will be set up for best-effort conversations. Allowable values are 16, 32, 64, 128, 256, 512, 1024, 2048 and 4096, with a default value of 256.

The third option, <reservable-queue> represents the number of reservable queues that will be used by RSVP conversations. The range of this option is 0 to 100, with the default value set to 0. If RSVP is enabled on a WFQ interface and the reservable-queues option is set to 0, the reservable queue size is automatically set by IOS to the interface bandwidth divided by 32 Kbps. Thus, if you require a different behavior you should set this option to a non-Ø value.

To better understand WFQ options let us focus our attention upon Figure 3.8 which illustrates the flow of data into queues. Note that the WFQ classifier automatically classifies frames by protocol, source and destination IP address, session layer protocol and source/destination port, IP precedence value in the Type of Service byte, and RSVP flow. The latter is a mnemonic for Resource Reservation Protocol, which represents a standard for allocating bandwidth across IP networks and will be described later in this book.

IP Precedence utilization

If we remember our discussion of IP Precedence earlier in this chapter when we examined the Type of Service field in the IPv4 header chapter, we noted that

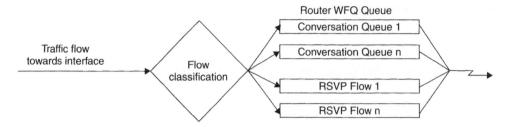

Figure 3.8 Weighted fair queuing

that field has values from 0 (the default) to 7. WFQ is IP Precedence aware and can detect higher priority market packets. In doing so the algorithm used by WFQ allocates additional bandwidth to conversations with higher precedence values. The allocation of additional bandwidth is based upon WFQ assigning a weight to each flow which is used to determine the transmit order for queued packets. Under WFQ lower weights are serviced before higher weights, with the IP Precedence functioning as a divisor in the weighting scheme used. For example, assume your router has one flow at each precedence level on an interface. Each flow is then assigned a precedence +1 value to remove the possibility of division by zero, resulting in the sum of flows becoming:

$$1 + 2 + 3 + 4 + 5 + 6 + 7 + 8 = 37$$

Thus, the flow with a precedence of 8 (actually a precedence bit value of 7) obtains 8/36ths of the bandwidth, while the flow with a precedence of 7 (corresponding to a bit value of 6) obtains 7/36ths of the bandwidth, and so on.

Now that we have an appreciation of the manner by which WFQ allocates bandwidth let us turn our attention to a variation of WFQ applicable referred to as Distributed-WFQ (D-WFQ).

Distributed-WFQ

Distributed-WFQ (D-WFQ) is only supported on certain Cisco Systems 7000 series routers that have either a Route Switch Processor (RSP), Versatile Interface Processor (VIP) or another type of enhanced processor. Under D-WFQ you can enable either flow-based queuing or class-based queuing by specifying the type of fair queuing to be placed into effect on an interface. To configure flow-based D-WFQ you would use one or more of the following command formats:

```
fair-queue
fair-queue aggregate-limit <aggregate-packet>
fair-queue individual-limit <individual-packet>
```

The first command is used to enable flow-based DWFQ. The second command, which is optional, sets the total number of buffered packets on the interface before some packets can be dropped. When under this limit, packets are not dropped. The third command is also optional. When used, this command option sets the maximum queue size for individual per-flow queues during periods of congestion.

Similar to WFQ, under D-WFQ when flow-based queuing occurs packets are classified by flow. Packets with the same metrics, such as source IP address, destination IP address, source TCP or UDP port, destination TCP or UDP port, and protocol are considered to belong to the same flow.

Options

In addition to flow-based queuing, under D-WFQ you can configure fair queuing based upon ToS or QoS-group. The format of the command to specify a specific type of fair-queuing is shown below:

`fair-queue [tos 1 qos-group]`

Using ToS results in ToS (precedence) based D-WFQ, while qos-group enables QoS-group-based D-WFQ. If you select tos you could then optionally specify the percentage of bandwidth to be allocated for each class. To do so you would enter the following command:

`fair-queue tos` <number> `weight` <number>

As another option you could set the number of buffered packets before some packets are dropped. To do so you would enter the following command:

`fair-queue aggregate-limit` <aggregate-packet>

In the preceding command the <aggregate-packet> number functions as a threshold. Below this number packets will not be dropped. A third option when fair-queuing based upon tos is used is the ability to set the maximum queue size for every per-flow queue during periods of congestion. To do so you would use the following fair-queue command format:

`fair-queue individual-limit` <individual-packet>

A fourth option provides you with the ability to set the maximum queue length for a specific ToS queue. To do so you would use the following fair-queue command format:

`fair-queue tos` <number> `limit` <class-packet>

Fair queuing based on ToS <number> varies between 0 and 7.

QoS-group

Under D-WFQ you can implement fair queuing based upon the configuration of QoS groups. To enable QoS-group-based D-WFQ you would enter the following command:

```
fair-queue qos-group
```

For each QoS group you need to specify the percentage of bandwidth to be allocated to each class. To do so you would use the following command format:

```
fair-queue qos-group <number> weight <weight>
```

Here the <number> variable would have the range 0 to 99.

Similar to our discussion of tos fair queuing, under QoS fair queuing you can optionally set the total number of buffered packets prior to some packets being dropped, the maximum queue length for every per-flow queue during periods of congestion, and the maximum queue size for a specific QoS group queue during periods of congestion.

To set the total number of buffered packets prior to some packets being capable of dropping you would use the following version of the fair-queue command:

```
fair-queue aggregate-limit <aggregate-packet>
```

In this command the <aggregate-packet> variable represents the number of packets prior to packets being dropped.

To set the maximum queue size for each per-flow queue during periods of congestion you would use the following optional fair-queue command:

```
fair-queue individual-limit <individual-packet>
```

The third option supported for the QoS-group enables you to set the maximum queue size for a specific QoS group queue. To do so you would enter the following command:

```
fair-queue qos-group <number> limit <class-packet>
```

Utilization examples

To illustrate a few examples of D-WFQ let us commence with a High Speed Serial Interface (hssi). To configure flow-based D-WFQ we would enter the following commands:

```
Interface hssi0/0/0
Ip address 192.72.46.1
Fair-queue
```

For a second example let us implement ToS-based D-WFQ. To do so on a hssi interface we would enter the following commands:

```
Interface hssi0/0/0
Fair-queue tos
```

For a third example we will configure QoS-group-based D-WFQ. In doing so we will implement two QoS groups, 2 and 6. Group 2 will correspond to an IP precedence of 2, while Group 6 will correspond to an IP precedence of 6. The following configuration sets up the previously mentioned QoS groups:

```
interface hssi0/0/0
ip address 198.78.46.1 255.255.255.0
rate-limit output access-group rate-limit 6
155000000 2000000 8000000 conform-action
set-qos-transmit 6 exceed-action drop
rate-limit output access-group rate-limit 2
155000000 200000 8000000 conform-action
set-qos-transmit 2 exceed-action drop
fair-queue qos-group
fair-queue qos-group 2 weight 10
fair-queue qos-group 2 limit 27
fair-queue qos-group 6 weight 30
fair-queue qos-group 6 limit 27
!
access-list rate-limit 2 2
access-list rate-limit 6 6
```

Rate-limit command

Because the rate-limit command was not previously covered let us discuss it now. The rate-limit command is used to configure committed access rate and distributed committed access rate policies. A committed access rate (CAR) both classifies packets and provides a mechanism that results in traffic flowing through an interface at or within a predefined rate. The latter function is also referred to as rate limiting. The format of the rate-limit interface command is shown below:

```
rate-limit [input/output] [access-group[rate-limit]
<access-limit-number><avg-bps><burst-bps-normal>
<burst-bps-max>conform-action<action>
exceed-action<action>
```

In examining the format for the rate-limit command you would use 'input' to apply the committed access rate policy to packets received on the interface while 'output' is used to apply the policy to packets transmitted on the interface. The keyword 'access-group' is optional and is used if you want to apply the policy to a specified access list. That list number follows the term 'rate-limit', which denotes a rate-limit access list. The next three variables denote the average bit rate in bps, the normal burst size in bytes, and the excess burst size in bytes. Thus, in the prior example, the average bit rate is 155 Mbps while the normal and excessive burst sizes are 2 Mbytes and 8 Mbytes, respectively.

Continuing our examination of the rate-limit command, the keyword 'con-form-action' is used to indicate an action should occur on packets that conform to the rate limit. Actions can include 'continue' which results in the next rate-limit command being evaluated, 'drop' which causes the packet to be dropped, 'set' which can include 'continue' to set IP precedence and evaluate the next rate-limit command or 'transmit' to set the IP precedence and transmit the packet. One additional action is 'transmit' which is used to indicate that the packet conforming to the rate limit should be transmitted. Last but not least the keyword 'exceed-action' is used to indicate the action to be taken on packets that exceed the rate limit. Those actions are the same as the actions noted for the 'conform-action' keyword.

Now that we have an appreciation of the operation of D-FWQ we will conclude our examination of Cisco router queuing methods with a discussion of a relatively new version of WFQ referred to as Class-Based WFQ.

3.8.3 Class-based weighted fair queuing

Class-Based Weighted Fair Queuing (CBWFQ) represents an additional queuing method introduced under IOS 12.1. CBWFQ represents a modified form of WFQ that operates on a class of traffic specified by the IP Precedence bits which until recently were known and referenced as the Type of Service (ToS) bits.

3.8.4 Functions to consider

Under CBWFQ you need to consider four general functions for each class of service you intend to have in your network. Those functions are classification, marking, policy and shaping and queuing. By permitting router administrators to use the command line interface (CLI) to set one item per command instead of multiple functions, the use of CBWFQ to configure QoS is both easier to understand and configure. To obtain an appreciation for CBWFQ let us focus our attention upon the four primary QoS functions performed in a router.

Classification

Classification represents the function that defines your policy and informs your routers how to recognize traffic belonging to a particular class of service. Classification can come into effect based upon criteria similar to previously described queuing methods. Such criteria can include the incoming interface of traffic, protocol, source or destination address, source or destination port or similar flow characteristics.

Marking

A typical result of the classification process is the marking of a packet. If queuing occurs only on an individual router basis marking may not be needed or, if required, occurs on a local basis. If traffic expediting is to occur on a network basis, a temporary tag can be added to packets. That tag performs two functions. First, the tag provides a mechanism for the routers within a network to appropriately route each packet. Second, by placing a tag at the beginning of a packet subsequent routers do not have to look deep into the packet to determine how to process the packet nor consult complex routing tables, which speeds up their packet processing capability. Later in this book we will turn our attention to Multi-Protocol Label Switching (MPLS) and how it uses a tag to expedite the flow of traffic through a network.

Policing and traffic shaping

Policing and traffic shaping can be considered to represent the manner by which communications devices can be programmed to react to market packets. For example, under policing, traffic up to a certain level is permitted. Thus, policing can be considered as a mechanism to enforce a service cap. In comparison, traffic shaping represents the manner by which packets are output via an interface. For example, packets could be output with a fixed or variable gap between each packet.

Queuing

Queuing represents the fourth method of providing QoS in a router or via a network. While classification and marking affect policing and traffic shaping, it is queuing which enables policing and shaping to be put into effect. To illustrate the use of CBWFQ let us first examine the commands needed to classify traffic.

3.8.5 Traffic classification

The class-map command is the key to creating a class map containing match criteria against which packets are checked to determine if they belong to a class. The format of this global configuration command is:

```
class-map<class-map-name>
```

Where the <class-map-name> variable represents the name of the class map to be created. Once the class-map command is used you need to specify one or more 'match' commands to denote what packets need to be matched to determine if they belong to a particular class. Table 3.3 lists the four match commands presently supported and a description of the use of each command.

To illustrate an example of the classification process let us assume we created an extended IP access list that was assigned the number 110. To define a class map that specifies that access list for packet checking to determine if they belong to the class we will label gold, we would use the following statements:

```
router(config) #class-map gold
router(config-cmap) #matchaccess-group 110
```

Table 3.3 Match commands associated with the class-map command

Match command	Description
`match access-group`[access-group| `name`<access-group name]	Specifies the name of the access list whose contents packets are checked to determine if they belong to the class
`match input-interface`<interface-name>	Specifies the interface used as a match criteria
`match protocol`<protocol>	Specifies the protocol used as a match criteria
`match mpls experimental`<number>	Specifies the value of the experimental field to be used as a match criteria

Table 3.4 Policy-map commands

Command	Description
`policy-map`<policy-map-name>	Specifies the name of the policy map
`class`<class>	Specifies the name of the class to be created and included in the service policy.
`bandwidth`[<bandwidth-kbps>| `percent`<percent>]	Specifies bandwidth on Kbps or percent of available bandwidth to be assigned to a class
`queue-limit`<number-of-packets>	Specifies maximum number of packets that can be queued for a class

Once you define your class map(s), the next step is to configure class policy using the policy-map command. You would use the policy-map command with one or more 'class', 'bandwidth' or 'queue-limit' commands to configure policy for each class. Table 3.4 lists the four relevant policy-map commands and a brief description of each.

The following snippet of code creates the policy named voip and allocates 10 percent of available bandwidth to the previously created class named gold.

```
router(config) #policy-map voip
router(config-pmap) #class gold
router(config-pmap) #bandwidth percent 10
```

Once you configure the class policy in the policy map a third required step is to attach the service policy and enable CBWFQ. To perform these two operations you would use the 'service-policy output' interface command. The format of this command is shown below:

`service-policy output` <policy-map>

where <policy-map> represents the name of the policy map.

In concluding our discussion of CBWFQ it should be mentioned that you can change the amount of bandwidth allocated for an existing policy map class, change the maximum number of packets that can be queued for an existing class as well as delete classes and policy maps. Because these functions are optional and our goal is to obtain an understanding of different Cisco queuing methods, we will not probe deeper into CBWFQ. Instead, readers who need to fully understand the latest developments in CBWFQ and other Cisco Systems queuing methods are cautioned to check the release of IOS they are using against the possibility that newer releases, such as 12.1, may support one or more queuing features not supported in the release they are using. If this situation occurs you may then have found a sufficient reason to upgrade to a newer release of IOS, assuming the additional queuing capability satisfies your organizational requirement.

4

DIFFSERV AND MPLS

In this chapter we will focus our attention upon a pair of standards recently developed to provide QoS in an IP environment – differentiated services (DiffServ) and Multi-Protocol Label Switching (MPLS). As we will note in this chapter, DiffServ takes the IPv4 Type of Service (ToS) field and modifies and reuses the contents of this field to transport information about IP packet servicing requirements. DiffServ operates at Layer 3 and makes no assumption about the underlying transport. In comparison, MPLS specifies how Layer 3 traffic can be mapped to such Layer 2 connection-oriented transports as frame relay and ATM. In doing so MPLS adds a label to each IP packet which enables routers to assign predefined paths to different traffic classes. Similar to prior chapters in this book we will first examine how a specific QoS technique operates. Once this is accomplished we will then turn our attention to implementing the technique in a Cisco environment.

4.1 DIFFERENTIATED SERVICES

The concept behind differentiated services (DiffServ) dates from the second half of 1997 when a number of proposals were made to the Internet Engineering Task Force (IETF). Those proposals were for a more scalable mechanism for differentiated services that would result in a number of well-defined traffic classes that could be used to prioritize the flow of data through a network. However, instead of having routers within a network noting flows, the resulting proposal that became DiffServ marks traffic that enters a network at the ingress node after checking for compliance against predefined service profiles. That marking is then used inside the network.

4.1.1 Ingress operations

Under DiffServ IPv4 and IPv6 packets can be labeled for a predefined QoS treatment. Under IPv4 the Type of Service byte is used while under IPv6 the Traffic Class byte is used. Readers are referred to Chapter 3, Section 3.1 for information on the type of Service and Traffic Class bytes under IPv4 and IPv6, respectively. Once a packet is marked Differentiated Services it is supported by routers within a DiffServ compatible network by queuing and forwarding packets based on the contents of the marked byte in the IP header.

4.1.2 Compatibility with IPv4 precedence

For both IPv4 and IPv6 the one byte ToS or Traffic Class fields are handled in a consistent manner under DiffServe. In doing so DiffServ provides backwards compatibility with the precedence portion of the IP ToS byte. As a refresher, Figure 4.1 illustrates the IPv4 Type of Service byte which will provide a frame of reference for noting how the DiffServ standard provides compatibility with the precedence bits of the ToS field.

Since the ToS field was proposed, its use was negligible until the explosive growth of the Internet resulted in congestion and the need to prioritize different types of traffic. Over the past five years different applications began marking the ToS field and router vendors provided different queuing methods that supported traffic expediting based upon the precedence bit settings in the ToS field. Due to this the IETF maintained compatibility with the precedence bits in the ToS field. However, because the other bits are not commonly used, the IEEE redefined the use of bits 3 through 5 to provide a total of 64 classes of service.

4.1.3 DiffServ bit labels

Under the DiffServ standard the most significant 3 bits (0, 1, and 2) are used for priority settings. However, DiffServ reorganizes and renames the precedence bits into a series of new categories. Those categories are listed in Table 4.1.

In examining the entries in Table 4.1 collectively, Classes 1 through 4 are referred to as Assured Forwarding (AF). As we will shortly note, under AF each class is provided with better or worse performance only in relationship to packets marked for the other AF classes.

```
 0   1   2   3   4   5   6   7
┌───────────────┬───┬───┬───┬───┬───┐
│  Precedence   │ D │ T │ R │ C │ 0 │
└───────────────┴───┴───┴───┴───┴───┘
```

Bits, 0, 1, 2 = Precedence   Bits 3, 4, 5, 6 7

Bits	Decimal	Precedence
111	7	Network Control
110	6	Internetwork Control
101	5	Critical
100	4	Flash Override
011	3	Flash
010	2	Immediate
001	1	Priority
000	0	Routine

Bit	Description	
3	Delay (D);	0 = Normal ; 1 = Low
4	Throughput (T) ;	0 = Normal ; 1 = High
5	Reliability (R) ;	0 = Normal ; 1 = High
6	Cost (C) ;	0 = Normal ; 1 = Low
7	Reserved ;	Set to 0

Figure 4.1 The IPv4 type of service field

Table 4.1 DiffServ precedence levels

Precedence level	Definition/Use
Precedence 7	Network Control
Precedence 6	Internetwork Control
Precedence 5	Express Forwarding
Precedence 4	Class 4
Precedence 3	Class 3
Precedence 2	Class 2
Precedence 1	Class 1
Precedence 0	Best Effort

4.1.4 Code points

To understand how the four classes of precedence are used under DiffServ requires us to examine bits 3 and 4 of the DiffServ byte. Those bits are now referred to as the Differentiated Service Code Points (DSCP) and are used in conjunction with bits 0 through 3 to provide a specification that denotes the probability associated with dropping a packet within a particular class. Under DiffServ the bit values of bits 3 and 4 are used to indicate low, medium and high packet dropping probabilities. Table 4.2 indicates the DSCP coding to specify the priority level (class) of a packet and its drop percentage. Note that bit 5 is always off under this scheme.

Although bits 3 and 4 are the actual codepoint bits, because they are used in conjunction with bits 0 through 2 the sequence of bits 0 through 5 (remember bit 5 is currently always set to 0) provides a 6-bit codepoint that theoretically can define 64 different classes of traffic. The 64 theoretical classes of traffic are subdivided into three pools of codepoints.

Codepoints of the form xxxxx0, where x is either a binary 0 or 1 are reserved for assignment as standards. Codepoints of the form xxxx11 are reserved for experimental or local use. The third pool of codepoints has the format xxxx01. This pool is also reserved for experimental or local use. However, this pool could be allocated for future standards if the need arises.

Codepoints of the form xxx000 are reserved to provide backward compatibility with routing based upon the Precedence field in the IPv4 ToS field. Another codepoint of interest is 000000, which represents the default packet class. Packets marked with this codepoint are treated on a best-effort forwarding basis. That is, such packets are forwarded on a FIFO basis as soon as bandwidth is available and higher priority packets have been output.

Table 4.2 DSCP coding by class

	Class 1	Class 2	Class 3	Class 4
Low drop percentage	001010	010010	011010	100010
Medium drop percentage	001100	010100	011100	100100
High drop percentage	001110	010110	011110	100110

Under DiffServ a compliant device first prioritizes traffic by class. Once this is accomplished, the device would differentiate and prioritize same-class traffic by examining the DSCP bit values that indicate the drop percentage. Collectively, as previously noted, Classes 1 to 4 are referred to as Assured Forwarding (AF). Under AF each class is provided with better or worse performance only in relationship to packets in the other classes. This means that if a link is congested, performance will be adversely effected in all classes. In comparison, Express Forwarding provides a mechanism for a service provider or an organization operating an extensive IP network to guarantee a minimum service level, since a minimum amount of bandwidth can be reserved for EF marked packets. In providing a minimum service level, packets marked as EF will receive the lowest level of latency and jitter.

4.1.5 Domain operations

DiffServ must be supported by all routers within a domain to be effective. Within a domain routers can be classified as either a boundary node or interior node devices as illustrated in Figure 4.2.

Boundary node functions

Boundary node DS compatible routers are responsible for five key functions – packet classification, packet metering, packet marking, packet shaping and packet dropping.

Packet classification

The classifier is responsible for separating packets into different classes according to a predefined set of rules. You can view the classifier function as a 1:N

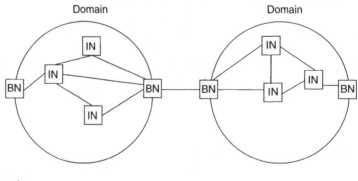

Legend :
 BN   boundary node
 IN   interior node

Figure 4.2 Boundary node and interior node routers within two DS domains

fanout device, taking a single traffic stream as input and generating N logically separate traffic streams as output. The classifier can be relatively simple and use only the DSCP in the IP header to determine the logical output stream to which the packet should be directed or it can be based upon multiple fields, such as the destination and source IP addresses, IP protocol, source destination TCP or UDP ports and the DSCP.

Packet metering

As the packets are classified the rate of packet flow must be compared to a threshold, a function referred to as metering. Metering also represents a logical 1:N fanout activity or function with statistics computed on the rate of packet flow compared to a predefined threshold or level.

Packet marking and dropping

If packets exceed a predefined level they can be either re-marked with a different codepoint or dropped. In the former situation, assume a given throughput is guaranteed for a particular service class. Any packets in that class that exceed the throughput level while bandwidth is still available could be re-marked for best effort handling. Another example of re-marking is when packets flow between two DS domains. Each domain could support different priority values or it is possible that one domain may not even support the DS markings of another domain unless an agreement exists between the operator of each domain. In comparison, if bandwidth is not available when the packet rate for a given class exceeds the profile for the class, the packet will be dropped. In actuality, under the DiffServ standards there are three levels of conformance defined with respect to metering and the comparison of the rate of packet flow to a given threshold. Those levels of conformance are 'colorful' as they are referred to as green (conforming), yellow (partially conforming) and red (non-conforming). The three levels can be used as triggers to invoke different queuing, marking or packet dropping treatments.

Packet shaping

The fifth function, packet shaping, can be considered to represent the delaying of packets. That delay ensures that the output rate for a given class does not exceed the traffic rate specified in the profile for a particular traffic class.

Core network functions

In comparison to a boundary node that performs classification, metering, marking, shaping and dropping routers within the core of a network handle packets in different traffic streams by forwarding them with different per hop behaviors (PHBs). The per hop behavior to be applied to a packet is based upon the value of the DSCP in the IP header of each packet.

Now that we have an appreciation of the concept behind DiffServ let us turn our attention to how DiffServ is supported on Cisco routers and switches.

4.2 SUPPORTING DIFFSERV IN A CISCO ENVIRONMENT

There are several methods that can be used to support DiffServ in a Cisco equipment environment. Two of the primary methods are through the use of Class Based Weighted Fair Queuing (CBWFQ) and Weighted Random Early Detection (WRED). Because we covered CBWFQ in Chapter 3 when we discussed Cisco queuing methods, we will briefly illustrate its use in conjunction with DSCP values in this section. Because WRED represents a congestion avoidance technique and was not previously described in this book, we will examine both how WRED operates and its use to support DiffServ in this section.

4.2.1 Using CBWFQ

The key to the use of Class Based Weighted Fair Queuing is to recognize that the recommended DSCP value for Expedited Forwarding is 101110 or decimal 46, while Assured Forwarding (AF) classes can be denoted as previously indicated in Table 4.2. In examining the entries in Table 4.2 note that AF class n where n varies from 1 to 4 is denoted by the DSCP value of xyzab0, where n has the value of xyz and ab represents the dorp precedence (dP) value. When you use CBWFQ you need to first determine the traffic classes and associated traffic classes and associated traffic types that correspond to each class. Next, you need to associate DSCP values to each traffic type in each class. To facilitate doing so this author converted the AF DSCP binary values to decimal, which are listed in Table 4.3.

4.2.2 Illustrative example

To provide a frame of reference we need an example. Let us assume we have four traffic classes that we will refer to as premium, gold, silver and bronze, with the latter three mimicking the series of Olympic medals. We will consider voice to be in the premium traffic class category. From Table 4.3 we will assume that all traffic marked with DSCP values of 10 and 12 corresponding to AF Class

Table 4.3 DSCP values for each AF per hop behavior

Drop precedence	Class 1	Class 2	Class 3	Class 4
Low	10	18	26	34
Medium	12	20	28	36
High	14	22	30	38

1 will be associated with traffic class gold. Traffic in the gold traffic class will include SNMP and TACACS. The silver traffic class will consist of Telnet, SMTP and FTP traffic. Such traffic will be marked with DSCP values of 18, 20 and 22 and represent the AF Class 2 category of traffic. Last but not least, the bronze traffic class will consist of HTTP traffic. We will assume HTTP traffic represents AF Class 3 traffic with a low drop precedence. Thus, HTTP will represent traffic marked with a DSCP value of 26. All other traffic will be considered to belong to a 'best-effort' traffic class. As mentioned in the previous section of this chapter, best-effort traffic has a DSCP value of 0. Table 4.4 lists the DSCP values that correspond to the previously mentioned traffic classes and traffic types.

Configuring CBWFQ

If we remember our prior discussion of CBWFQ, the configuration process requires us to define class maps and configure a class policy in the policy map. Once this is accomplished we need to attach a service policy and enable CBWFQ. Optionally, we can modify the bandwidth for an existing policy map class, modify the queue limit for an existing policy map class, and perform such other functions as changing the maximum reserved bandwidth allocated to CBWFQ and deleting classes and policy maps previously created. Let us assume traffic will flow from a Fast Ethernet LAN switch into a router connected to a TCP/IP network. Thus, we will need to configure a policy map and create class policies that make up the service policy previously mentioned and listed in Table 4.4 for the router's Fast Ethernet port. To do so we would configure the policy map contained in Figure 4.3, which we will name SETDSCP.

In examining the entries in the SETDSCP policy map shown in Figure 4.3 note that the 'class-map' command is used to define a traffic class. Each traffic class has three major elements – a name, one or more 'match' commands, and an instruction which denotes how to evaluate the 'match' command(s). Thus, all packets are first matched to determine if they have a DSCP value of decimal 46. The reason that packets are first checked for a DSCP value of 46 results from the 'policy-map' portion of Figure 4.3. Note that after the 'policy-map' command sets the name of the policy to SETDSCP, the first 'class' command specifies the class to be created as EF. This is followed by the use

Table 4.4 DSCP values to be used for an example of traffic classes and traffic types

Traffic class	Traffic type	DSCP value
Premium	Voice over IP	46
Gold	SNMP	10
	TACACS	12
Silver	Telnet	18
	SMTP	20
	FTP	22
Bronze	HTTP	26
Default	All other traffic	0

```
class-map match-all EF
 match access-group 101
 !
class-map match-all AF11
 match access-group 102
 !
class-map match-all AF12
 match access-group 103
 !
class-map match-all AF21
 match access-group 104
 !
class-map match-all AF22
 match access-group 105
 !
class-map match-all AF23
 match access-group 106
 !
class-map match-all AF31
 match access-group 107
 !
policy-map SETDSCP
 class EF
  set ip dscp 46
 class AF11
  set ip dscp 10
 class AF12
  set ip dscp 12
 class AF21
  set ip dscp 18
 class AF22
  set ip dscp 20
 class AF23
  set ip dscp 22
 class AF31
  set ip dscp 26
```

Figure 4.3 The SETDSCP policy map

of the 'set ip dscp' command that specifies the value of the IP DSCP field to be associated with the previously defined class named EF. Similarly, the other classes required are also defined in the lower portion of Figure 4.3. Returning to the top portion of Figure 4.3, note that for each 'class-map' command a 'match' command is used to define a particular match criteria. That criteria will be defined by the use of an access-group. Thus, in effect the first 'class-map' command defines the traffic class EF that has a DSCP value of 46 and will consist of packets that match the criteria in access-group 101.

Another item to note concerning the entries in Figure 4.3 is the numbering convention use for Assured Forwarding (AF). The first digit following AF represents the class while the second digit represents the drop precedence within a class, with 1 being low, 2 representing medium, and 3 denoting high.

Defining traffic class behavior

Figure 4.3 listed the commands necessary to mark each traffic class with appropriate DSCP values. Once that task is accomplished we need to define the different behavior aggregate requirements for each of the traffic classes.

Because voice over IP is considered as the premium class, we will want to forward traffic with a DSCP value of 46 with the lowest possible delay. Let us assume premium class should obtain up to 784 Kbps of bandwidth on the router's serial port during periods of congestion. Let us also assume that the gold, silver, and bronze classes require 20, 15, and 10 percent, respectively, of the serial interface bandwidth as their minimum bandwidth guarantees. Finally, we will shape silver and bronze traffic classes to 256 Kbps and 28 Kbps, while the best effort traffic class should be policed to 56 Kbps. To configure the previously mentioned settings on the serial port of the router we would again create a series of class maps and policy map. However, this series of class maps and the resulting policy map would be applied to the router's serial port in the outbound direction. Figure 4.4 lists the applicable commands for creating the policy map we will name SERIALOUT.

In examining the entries in Figure 4.4 for the policy map labeled SERIALOUT the 'class-map' command is again used to define different traffic classes. As previously noted, each traffic class has three major elements – a name, one or more 'match commands and an instruction which denotes how to evaluate the 'match' command(s).

```
class-map match-all premium
 match ip dscp 46
class-map match-all gold
 match ip dscp 10 12
class-map match-all silver
 match ip dscp 18 20 22
class-map match-all bronze
 match ip dscp 26
class-map best effort
 match access-group 110
!
policy-map serialout
 class premium
  priority 784
 class gold
  bandwidth percent 20
 class silver
  shape average 256000
  bandwidth percent 15
 class bronze
  bandwidth percent 10
 class best effort
  policy 56000 1550 1550 set-dscp-transmit 0 ie:-transmit 0
```

Figure 4.4 The creation of a policy map and class maps that define a policy on the serial interface

Using the priority command

In Figure 4.4 we first match all packets against a dscp value of 46. Under the policy map portion of the listing shown in Figure 4.4 the 'priority' command is used to guarantee 784 Kbps of bandwidth to the premium class of traffic. Since premium traffic must have a DSCP value of 46 and that value represents EF traffic, in effect we are guaranteeing 784 Kbps of bandwidth to EF traffic.

As a refresher, the format of the 'priority' command is shown below:

```
priority [<kbps>] percent<percent>] [<bytes>]
```

Note that you can guarantee allowed bandwidth by either specifying a data rate in kbps or a percentage of the transmission capacity. As an option, the <bytes> argument controls the size of the burst allowed to pass through the device without being considered in excess of the configured kbps rate.

Using the bandwidth command

In comparison to the use of the 'priority' command that guarantees a specified amount of bandwidth, the 'bandwidth' command is used to specify a minimum amount of capacity or bandwidth for a traffic class. While this difference may appear to be one of semantics and equivalent to splitting hairs, it is not. That is, the 'priority' command guarantees a specific amount of bandwidth while the 'bandwidth' command allocates a minimum amount of bandwidth. Thus, there is a distinct difference between the two commands.

The format of the 'bandwidth' command is shown below:

```
bandwidth [<kbps>] percent<percent>]
```

Similar to the 'priority' command, you can specify the minimum bandwidth in the 'bandwidth' command either in terms of Kbps or as a percentage of the overall available bandwidth. Thus, returning our attention to Figure 4.4, we used the bandwidth command to allocate a minimum of 20 percent of available to the gold traffic class, 15 percent to silver and 10 percent to bronze.

Two additional commands used in the listing shown in Figure 4.4 that require a bit of elaboration are the 'shape' and 'police' commands so let us focus our attention upon each.

Using the shape command

The 'shape' command is used to shape traffic to an indicated bit rate. Traffic shaping will occur according to the variable specified that defines a particular shaping method. The format of the command is shown below:

```
Shape [average 1 peak ]<mean-rate> [<burst-
size>[<excess-burst-size>]]
```

Returning our attention to the listing contained in Figure 4.4, the 'shape' command shapes traffic to an average data rate of 256 Kbps for the silver traffic class. However, the 'bandwidth' command insures that the silver traffic class always has a minimum of 15 percent of the bandwidth. If the serial port was a T1 line operating at 1.544 Mbps, the use of the 'shape' and the 'bandwidth' commands are equivalent stating that the silver class traffic should have a minimum of 1.544 Mbps × .15 or 231.6 Kbps and an average data rate of 256 Kbps. Thus, silver traffic will not be very variable.

Using the police command

The 'police' command can be viewed as placing a cap on the use of bandwidth. This command is used to specify a maximum amount of bandwidth available for a specific traffic class. The format of this command is indicated below:

```
police<bps><burst-normal><burst-max>[conform-action
<action>1exceed-action<action>1violate-action<action>]
```

The specification of a maximum bandwidth for use by a particular traffic class occurs through the use of a token bucket algorithm. Thus, prior to focusing our attention on the use of the 'police' command in the listing shown in Figure 4.4 let us review what is meant by a token bucket algorithm.

Token bucket algorithm

A token bucket represents a mechanism to define a traffic profile which takes into consideration the packet rate and burstiness. The term bucket represents a counter that indicates the allowable number of bytes that can be transmitted at a particular time. Graphically, a token bucket operation is illustrated in Figure 4.5.

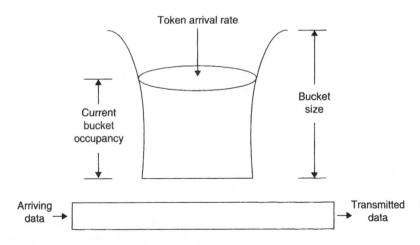

Figure 4.5 A token bucket regulates traffic flow

In this example tokens arrive at an arrival rate (AR) up to the capacity of the bucket. The arrival rate causes the counter to increment up to the maximum counter value which is equivalent to the bucket capacity. As packets arrive, tokens that are metaphorly placed into the bucket are extracted and serve as a flag to permit packets to be forwarded. If the bucket fills to capacity, newly arriving tokens are discarded. You can view each token as permission to obtain the ability to transmit a certain number of bits into the network. As packets are transmitted a certain number of tokens will be removed from the bucket.

If there are not enough tokens in the bucket to transmit a packet, the transmitter must either wait until the bucket has a sufficient number of incoming tokens or discard the packet. In comparison, if the bucket is full of tokens, incoming ones overflow and cannot be used to service future packets. Thus, at any point in time the largest burst a transmitter can send into the network is approximately equivalent to the capacity of the bucket.

For traffic shaping a token bucket has three components: a burst size, a mean rate and a time interval. The mean rate is normally represented in terms of bps and the relationship between the three is as follows:

```
mean rate = burst size
            ----------
            time interval
```

Now that we have an appreciation for the function of a token bucket let us turn our attention back to the 'police' statement. The `<bps>` variable indicates the average rate in bps which in our example is 56 Kbps and was entered in the listing in Figure 4.4 as 56000. The `<burst-normal>` variable denotes the normal burst size in bytes. In Figure 4.4 the entry was 1550 which we will assume is of sufficient length to forward a maximum length Ethernet frame that is converted to a packet by the router. The next variable is `<burst-max>` which represents the excess burst size in bytes. As a notation and also as a reminder to check the release of IOS you are working with, this variable is optional in IOS release 12.1(5)T onward, unless the violate-action option is specified. Under IOS Release 12.0(5) × E through 12.1(1)E the excess burst size must be specified.

The conform-action keyword in the command line is used to define the action that will occur on packets that conform to the rate limit. In comparison, the keyword 'exceed-action' denotes the action that will occur on packets that exceed the rate limit. The third action, specified by the keyword 'violate-action' denotes the action that will be taken on packets that violate both the normal and maximum burst sizes. If you specify this option, the token bucket algorithm is put into effect using two token buckets.

For all three keywords there are five options that can be set through the use of predefined keywords. Those action keywords and their meanings are listed in Table 4.5.

Returning once again to the listing in Figure 4.4, the last line in the listing results in packets that conform to the rate limit having their DSCP value being set to 0.

Table 4.5 Police command action keywords

Keyword	Meaning
drop	Drop the packet
set-prec-transmit<precedence>	Sets the IP precedence value and transmits the packet
set-qos-transmit<new-qos>	Sets the QoS group and transmits the packet
set-dscp-transmit<dscp-value>	Sets the DSCP value and transmits the packet
transmit	Transmits the packet

Setting the access lists

As we continue our examination of modular snippits of IOS commands let us now turn our attention to the relevant access-lists commands necessary to define the traffic types previously listed in Table 4.4 and the access groups defined in the listing contained in Figure 4.4. Figure 4.6 contains the series of access-list statements that define the type of traffic we will work with.

In examining the access-list statements shown in Figure 4.6, while most statements are self-explanatory, a few of deserve a degree of elaboration. First, access- list 101 which corresponds to Expedited Forwarding (EF) for a VoIP application defines UDP ports within the range 16384 through 32768. That range is commonly used by Cisco VoIP products. However, if you are using a different vendor's product you should check the port or port values to ensure an appropriate access-list configuration. A second item that warrants a few words of elaboration is the use of the keyword 'any' for both source and destination addresses for each access-list statement. For this example it was assumed that all source and destination addresses were applicable, however, you would want to consider applicable address restrictions to satisfy your organization's operating environment. One additional item prior to moving on is the last access-list statement. That statement corresponds to best-effort traffic and represents traffic that is not categorized as premium, gold, silver or bronze.

Now that we have an appreciation of how we can configure CBWFQ let us turn our attention to a second method that can be used with DiffServ. That method

```
access-list 101 permit udp any any range 15384 32768
access-list 102 permit udp any any eq snmp
access-list 103 permit tcp any any eq tacacs
access-list 104 permit tcp any any eq telnet
access-list 105 permit tcp any any eq smtp
access-list 106 permit tcp any any eq ftp
access-list 107 permit tcp any any eq http
access-list 108 permit ip any any
```

Figure 4.6 Access-list statements for the application developed in this section

is distributed weighted random early detection (DWRED), which represents a congestion avoidance mechanism.

Using DWRED

To obtain an appreciation of the manner by which distributed weighted random early detection (DWRED) operates with DiffServ requires an understanding of the manner by which weighted random early detection (WRED) operates. Because WRED represents a congestion avoidance technique that operates by managing queue depths, it is not a queuing technique. Thus, we did not cover WRED in Chapter 3 when we discussed different queuing techniques so let us do so now.

Weighted random early detection

Weighted random early detection (WRED) has its roots in random early detection (RED), a congestion avoidance technique developed during the 1980s to facilitate the flow of TCP/IP traffic. To obtain an appreciation of how RED operates we will once again digress a bit and review how TCP operates and how it includes a built-in algorithm that enables it to vary the size of its send window.

TCP operations

One of the built-in features of TCP that is transparent to users is the manner by which it varies the size of its transmit window. TCP measures the time intervals from the transmission of a segment of data until an acknowledgement is received from the destination station. This interval represents a round-trip delay to include network and host processing time. TCP uses this time to vary the size of its transmit window, since the size of that window governs the rate at which transmission occurs. That is, a long window permits more data to be transmitted prior to the receipt of an acknowledgement than a shorter window. Although under normal conditions the window size remains constant, minor variations in the round-trip delay result in TCP adjusting two threshold values. Those thresholds represent minimum and maximum round-trip delays. If the maximum threshold is exceeded prior to the receipt of an acknowledgement, TCP will assume a segment is lost and will retransmit the previously transmitted information. In addition, TCP will set its congestion window to a value of 1 segment and implement its slow-start transmission method. Under slow-start, TCP uses a congestion window, which is assumed and not included in the protocol header. As a new connection is established first, one segment is transmitted. Each time an ACK (acknowledgement) is received the congestion window is doubled. The transmitting station can transmit a number of segments up to the minimum of the value of the congestion window or the advertised window. This technique results in a dual flow control. That is, the sender is limited by the congestion window that it controls but, in addition, it is also limited by the advertised window, which is controlled by the destination.

The slow-start mechanism provides for congestion avoidance as it limits the data flow from the origination. Prior to slow-start being implemented, under TCP an initial connection between two stations could result in one station bursting multiple segments up to the advertised window size of the receiver established during the 3-way handshake. While this is usually not a problem when both stations are located on the same LAN, when located on different networks, start-up without slow-start could adversely effect the local router's ability to service a mixture of traffic in a timely manner.

Under slow-start the receipt of each ACK results in the sender doubling its congestion window. This doubling continues until the size of the congestion window equals the size of the advertised window. When this situation occurs, segments are transferred as if slow-start never occurred.

The congestion window can be both incremented and decremented. Concerning the latter, the window size is decreased upon congestion indicated either by a time out or the receipt of duplicate ACKs. When either situation occurs a comparison is made between the congestion window size and the advertised window size. Whichever is smaller has its value halved. That value is then saved as a slow-start threshold variable. That value must be at least 2 segments unless the congestion occurred due to a time-out, with the latter causing the congestion window to be set to 1. At this point in time the TCP sender can either start up in slow-start mode or in congestion-avoidance mode. As ACKs are received, the congestion window is increased. When its value matches the value saved in the slow-start threshold variable, the slow-start algorithm will stop and the congestion-avoidance algorithm will commence, resulting in a more linear growth in transmission instead of the exponential growth of the slow-start algorithm.

Congestion collapse

Although TCP's built-in slow-start and congestion avoidance capability reduces the initial data flow of a session, it operates on a session-by-session basis. If you have n senders that begin transmission at times $t_0, t_1, t_2, t_3 \ldots t_n$, with each interval very close to one another, it becomes possible for cumulative traffic to overwhelm the transmission capacity of a circuit. When this situation occurs router buffers fill until they overflow, resulting in packets becoming delayed or lost. In addition, retransmission on a LAN occurring at a higher data rate than most WAN operating rates result in the rapid filling of router buffers and become responsible for additional traffic being dropped. The result of packet dropping from numerous TCP sessions commencing close to one another represents the primary cause of a sequence of a traffic build-up followed by a collapse of traffic that resembles an oscillation that repeats on a periodic basis. Figure 4.7 illustrates the effect of a series of TCP sessions cycling through congestion build-up and collapse over a period of time.

Now that we have an appreciation of the manner by which TCP sessions can result in a sequence of congestion and traffic collapse, let us turn our attention to the manner by which random early detection counters this problem.

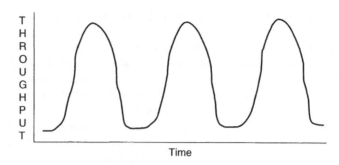

Figure 4.7 TCP congestion build-up and collapse

Random early detection

Random early detection (RED) represents a queue manager software module that can be used to limit the effect of congestion build-up. RED monitors the depth of output queues and randomly drops packets in order to limit the depth of queues. In actuality, RED uses an algorithm which computes the probability for dropping a packet as well as the exponential weighing factor for determining an average queue size. The latter is used for comparison against a threshold to determine if the packet is queued or dropped.

The average queue size is based upon the previous average and the current size of the queue. The formula for the average queue size is:

```
Avg_queue_size=(old_average*(11/2^n))+(current_queue_size*1/
2^n)
```

This is where n represents the exponential weight factor that can be configured by the user.

The probability that a packet will be dropped is based upon a minimum threshold value, maximum threshold value and mark probability denominator. When the average queue size exceeds the minimum threshold value, RED begins to drop packets, with the drop rate increasing linearly as the average queue size increases until it reaches the maximum threshold when all packets are dropped. The mark probability denominator is the fraction of packets dropped when the average queue size reaches the maximum threshold. For example, if the denominator is 256, one out of every 256 packets is dropped when the average queue size is at the maximum. Figure 4.8 illustrates the packet discard probability as a function of the average queue size.

Based upon the preceding we can summarize the use of RED upon packets as follows:

- When a packet arrives, the average queue size is computed.
- If the average queue size is less than the minimum queue threshold, the packet is queued.
- If the average queue size is between the minimum and maximum threshold, the packet is either dropped or queued depending upon its drop probability.

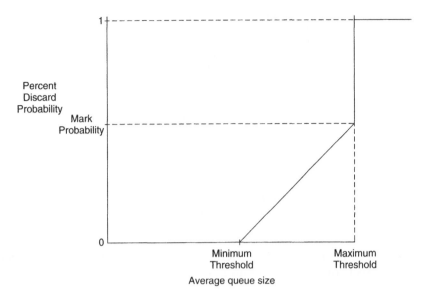

Figure 4.8 RED packet discard probability as a function of the average queue size

- If the average queue size is greater than the maximum queue threshold, the packet is dropped.

Weighted RED

Although RED works with TCP it does not work with UDP. Recognizing this limitation Cisco developed weighted RED (WRED) which queues traffic according to precedence values. WRED provides packet-dropping thresholds for different IP precedence values. This enables lower priority traffic to be dropped before higher priority values.

WRED maximum queue-depth thresholds are based upon the interface speed of the router port and the output buffers available for the particular interface. The default values for the minimum queue-size thresholds are based upon a fraction of the maximum threshold. Table 4.6 lists the default values for each IP precedence value.

Configuring WRED

The basic format of the command to enable WRED is shown below:

```
random-detect [exponential-weighting-constant n]
```

This is where n represents a number from 1 to 16. By default the exponential weighting factor is 9. The higher the value, the lower the responsiveness of RED to queue-depth changes, resulting in a smoother response. A second version of the random-detect interface is used to modify minimum and maximum

Table 4.6 WRED queue-size default threshold values as a fraction of the maximum threshold

IP precedence	Default minimum
0	9/18
1	10/18
2	11/18
3	12/18
4	13/18
5	14/18
6	15/18
7	16/18
RSVP	17/18

thresholds and a mark-probability-denominator. The format of this version of the random-detect interface command is shown below:

```
random-detect [precedence <p>] [<min-threshold>]
[<max-threshold>] [<mark-prob-denominator>]
```

where:

```
p = IP precedence values, 0-7 or rsvp.
Min-threshold = lowest queue size where random
discards begin; 1-4096
Max-threshold = largest queue size before packets
being dropped, values are from (min-threshold-4096)
Mark-prob-denominator = number of packets that can be
queued before discards occur. Values range from 1 to
65535.
```

Now that we have an appreciation of the operation of WRED it is time to focus our attention upon the use of differentiated services with WRED.

DiffServ and WRED

Support for DiffServ and WRED was introduced in Cisco IOS Release 12.1(5a)E and is also available in Release 12.0(15)S or later on VIP enabled Cisco 7500 routers and certain Catalyst switches. Officially referred to as DiffServ WRED, this feature enables DWRED to use the DSCP value when it calculates the drop probability for a packet. To accomplish this, the random-detect command adds two new arguments – 'dscp-based' and 'prec-based' to the random-detect policy map class command. The dscp-based argument enables DWRED to use the DSCP value of a packet when it computes the drop probability of a packet. In comparison, the 'prec-based' argument enables DWRED to use the IP Precedence value of a packet when it computes the drop probability for a packet.

```
Router(config-if) # class-map web
Router(config-cmap) # match access-group 101
!
Router(config-if) # policy-map pweb
Router(config-pmap) # class web
Router(config-pmap-c) # bandwidth 128
Router(config-pmap-c) # random-detect dscp-based
Router(config-pmap-c) # random-detect dscp 26 24 64
!
Router#interface s0
Routerconfig-if) # service-policy output s0
```

Figure 4.9 Configuring DWRED to use a DSCP value

Configuration

To configure DWRED to use the DSCP value when computing the packet drop probability you would create a traffic class using the 'class-map' command for matching packets to a particular class and use the 'policy-map' command to create a traffic policy. You would then either use the 'bandwidth' or 'shape' commands to set a minimum bandwidth guarantee or shape the traffic to an indicated bit rate. Once the prior operations are performed you would use the following command to inform DWRED to use the DSCP value when it computes the drop probability for a packet.

```
random-detect dscp-based
```

Next, you need to specify the minimum and maximum packet thresholds and optionally, the mark-probability denominator for the DSCP value. To do so you would use the following command:

```
random-detect dscp<dscp-value><min-threshold><max-
threshold>[<mark-probability-denominator>]
```

Last but not least, you would attach the traffic policy to an interface in the output direction. To do so you would use the following command:

```
service-policy output <policy-map>
```

To illustrate an example of the configuration of DWRED to use DSCP values, let us assume we want to create a class named web on which the DSCP value of 26 will be used by DWRED. We will set a minimum and maximum packet threshold of 24 and 64 and will apply the traffic policy to all traffic leaving interface sØ. To do so we would use the commands listed in Figure 4.9.

4.3 MULTI-PROTOCOL LABEL SWITCHING

Multi-Protocol Label Switching (MPLS) represents a relatively new, standards-based protocol from the Internet Engineering Task Force (IETF) that enables

Layer 3 traffic to be mapped to connection-oriented Layer 2 transported, such as frame relay and asynchronous transfer mode (ATM). Because frame relay and ATM can be engineered to provide a QoS traffic capability, the use of MPLS means that it becomes possible to use those Layer 2 technologies to provide IP traffic with a QoS capability. As we move forward into the new millennium MPLS is even being considered as a mechanism to mate IP directly onto optical networks, bypassing the complexity of the Synchronous Optical Network (SONET). Of course, you need to have the underlying Layer 2 transport to do so.

In this section we will focus our attention upon obtaining an appreciation of how MPLS operates. In doing so we will note how a label switching router (LSR) at the edge of a network assigns labels to each incoming packet and how such labels are used for routing traffic over a Layer 2 network.

4.3.1 Overview

MPLS evolved from two similar technologies referred to as IP switching from Ipsilon Networks and tag switching developed by Cisco Systems. Under the IETF standard MPLS defines both an architecture as well as a protocol for encapsulating IP traffic within new routing headers.

Rationale

The rationale behind MPLS is to mate connectionless IP onto a Layer 2 connection-oriented transport. This enables IP traffic to be regulated to the point where a QoS capability can be provided. In addition, by prefixing a label to each packet that indicates its flow through a network, the complexity associated with IP routing is avoided, enhancing the flow and processing of IP packets.

Restrictions

Because MPLS is an edge-to-core technology it is not a feature that should be considered by small organizations. Instead, MPLS is oriented for use by service providers as well as large organizations that operate an enterprise-wide infrastructure.

Traffic flow

Figure 4.10 illustrates the flow of traffic from a non-MPLS network into and through an MPLS network. Moving from left to right in Figure 4.10, at the edge of the MPLS network a label edge router (LER) adds a 4-byte label to each IP packet. The purpose of the label is to enable routers within the core of the Layer

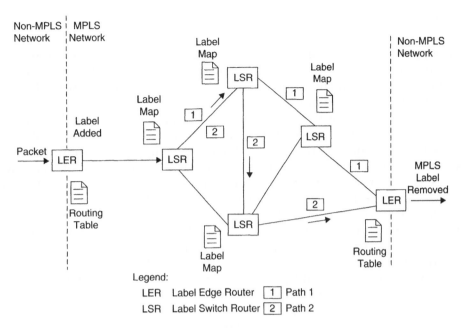

Figure 4.10 Traffic flow through an MPLS network

2 network to forward packets using predetermined paths based upon specified QoS levels.

The label added at the edge of the network informs the next-hop MPLS compliant router of either the packet's predefined path or a set of predefined paths that the packet can travel. As the packet flows across the Layer 2 network it may be relabeled to travel a more efficient path. Upon exiting the MPLS network, the label is removed from the header which results in the packet being restored to its original size.

Obtaining QoS

In examining the flow of traffic into and through the MPLS network those readers familiar with the capability of ATM and frame relay can understand how the Layer 2 network can provide a predefined QoS. For those not familiar with the operation of those networks let me briefly digress to explain how QoS becomes possible. In an ATM environment a constant bit rate (CBR) in effect provides a guarantee of packet (in the form of converted data being transported as cells) delivery with constraints governing latency between cells. In a frame relay environment a service level agreement (SLA) will normally guarantee 99.9 percent or more of all frames flowing up to a committed information rate (CIR) for delivery to their destination. Thus, the trick to obtaining an IP QoS capability is to map packets into an applicable stream of ATM cells or frame relay frames. However, what may not be clear is how a label edge router at the IP network to MPLS network boundary knows the type of service level desired. The answer to this question is via an examination of the IP header.

Label processing

When a packet enters an MPLS network, route information is bound to a label affixed to the packet. Within the MPLS network each node forwards packets with the same label in the same manner. That is, two packets with the same label would be forwarded out the same port.

In actuality there are different types of information that can be conveyed within a packet label. Labels can contain a host route, a combination of source and destination information, or be used to denote packets that share a common class of service or even equate to a particular organization's virtual private network (VPN). In fact, the use of MPLS to support VPNs has significantly increased over the past few years. The reason for the growing acceptance of MPLS for VPNs results from the fact that its use provides more flexibility than an Internet Service Provider building VPNs directly over ATM or frame relay permanent virtual circuits (PVCs) or creating different types of tunnels to interconnect customer routers located at the edge of Layer 2 networks. When VPNs are created using MPLS, customers can select their own addressing plan. Because the label process as we will shortly note restricts the flow of packets within the MPLS network so that they can only exit to customer VPN sites, the customer may elect to do without encryption and authentication, two necessary functions when creating a VPN directly over IP. In addition, because the customer only needs to configure routers located at their premises for a static route to the service provider, the use of an MPLS VPN operated by a service provider can also significantly reduce potential routing complexities.

Returning our attention to Figure 4.10, when a packet enters an MPLS-compliant network the LER will examine the packet's IP header in order to classify it and assign a label. While the packet will be forwarded using normal Layer 3 techniques to the next hop, at that hop the MPLS node performs a simple function in place of a router table look-up. The MPLS node looks up the label in a table, swaps labels, may decrement the IP time-to-live (TTL) field and forwards the packet.

At each MPLS node a label has only local significance. That is, at each hop a router will strip off the current label and add a new one, forwarding the packet based upon the information in the new label similar to the manner by which cells and frames are forwarded in ATM and frame relay network since those Layer 2 networks typically form the foundation for the MPLS network. In fact, in an MPLS/ATM environment the MPLS label is the Virtual Channel Identifier (VCI)/Virtual Path Identifier (VPI) of an ATM virtual connection. In a frame relay network the Data Link Control Identifier (DLCI) is the label. Thus, IP switching occurs at the speed of the underlying ATM or Frame relay switching system. Because labels are switched at each node within an MPLS network many times, an MPLS network is referred to as a label switching network. At each node or hop within the network packets are routed based upon the value of the incoming label and forwarded through a different interface with a new label. The path that a packet flows through an MPLS network is defined by the sequence of label values as labels are removed and added to each packet as they flow through each node in the network. Because the mapping between labels is constant at each label switch router, the path is determined by the initial label value placed on the packet at the edge of the network. The path is known as a label switched path (LSP). A flow of packets determined by a predefined criteria, such as precedence value or protocol, is referred to as a Forwarding Equivalence Class (FEC) as all such packets are treated the same. One or more FECs can be mapped to a single LSP to provide a similar QoS to multiple traffic flows, however, network engineers must ensure there is sufficient capacity through the initial LER and the LSRs in the network to provide the needed QoS.

Hardware

In an ATM environment the edge of the network commonly uses an MPLS router with ATM switching functionality. Network engineers must configure a virtual circuit through the network. In addition, ATM switches in the core of the network must have MPLS capability as a label switching function is required to facilitate the movement of IP traffic transported on ATM cells through the network.

4.3.2 Operational example

To illustrate the use of labels and the label swapping process within an MPLS network let us examine a small portion of an MPLS network. Figure 4.11 illustrates four routers within an MPLS network.

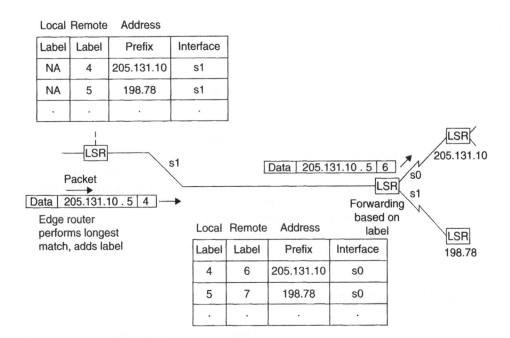

Figure 4.11 MPLS routing and label swapping example

The router located at the extreme left portion of Figure 4.11 represents an edge router which applies an initial label to each packet entering the MPLS network after performing a conventional longest-match look-up on the destination address in the IP header. In this example, after performing a longest match on the destination IP address in the header, the edge router affixes a label to the packet. Based upon the contents of the label table shown in the upper left corner of Figure 4.11 the label added has the value 4 and is output on serial interface 1. Upon receipt of the label at the first label switch router within the MPLS network, the router examines the label, notes its value of 4, examines its label table, and swaps labels, prefixing a label with a value of 6 onto the packet and then forwarding the packet out of the router's sØ port. Note that within the MPLS network all packet forwarding operations occur based upon the value of the label of the arriving packet and the contents of the label table at each LSR that services the packet on its route to its destination.

4.3.3 MPLS QoS

In a Cisco equipment environment a distinction is made between edge and core network devices when providing MPLS QoS. Because QoS requires intensive processing the Cisco model depends upon edge devices to perform most of the processor-intensive work to include required flow identification, packet classification, and bandwidth management. In comparison, core devices are responsible for expedited forwarding using labels while enforcing QoS levels assigned at the edge of the network. Under a Cisco equipment environment you can construct an end-to-end QoS architecture in several ways. First, you can use IP precedence bit values in the IP header to indicate the service class of a packet. The value set at the edge will be enforced in the core of the network, with different label values used to indicate different precedence levels.

A second method available is to use a committed access rate (CAR) under which multiple Layer 3 thresholds would be configured based upon such parameters as the application transported or protocol. If the flow exceeds a predefined threshold, different responses would occur, such as dropping packets or sending them at a lower service class.

Two additional Cisco QoS features that can be used at the edge of a network include WRED and CBWFQ. Under WRED network congestion is avoided by slowing flows based upon service class. In comparison, under CBWFQ, different weights are assigned to different traffic classes and packets can be reordered to control latency both at the edge as well as in the core of the network. Because weights are relative and not absolute this means that bandwidth of unfertilized resources can be shared between other service classes.

4.4 CONFIGURING MPLS

In this section we will turn our attention to examining the tasks required to configure MPLS in a Cisco environment. However, prior to doing so it should be noted that similar to many other QoS features, MPLS is not supported on all Cisco products. Thus, readers should examine the latest Cisco literature to determine the hardware platforms supported. In addition, the ability to use MPLS depends upon the type of IOS services your organization is using. For the example we will use, which is to use MPLS to create virtual private networks, you need to run MPLS in provider backbone routers, MPLS with VPN code in provider routers with VPN edge service (PE) routers, Border Gateway Protocol (BGP) in all routers providing a VPN service, and Cisco Express Forwarding (CEF) in each MPLS enabled router.

Concerning the enabling of MPLS, the process required to turn it on is relatively simple. If your router supports distributed Cisco Express Forwarding (DCEF) you would enter the following command:

```
ip cef distributed
```

To turn on MPLS tag distribution you would enter the following command:

```
tag-switching advertise-tags
```

A third step you need to perform is to enable MPLS on an applicable interface or set of interfaces. For example, if MPLS is to be set on the serial Ø interface, you would enter the following commands:

```
interface sØ
tag-switching ip
```

4.4.1 Configuration tasks

There are five mandatory tasks required to configure VPNs. In addition, it is always a good idea to verify your configuration through the use of applicable 'show' commands. Thus, we can say that there are six key tasks required to configure and verify VPNs. Those tasks are listed in Table 4.7.

4.4.2 Network configuration

Since the best way to obtain an appreciation of creating an MPLS based VPN is by examining a network configuration, let us do so. We will literally look at the

Table 4.7 VPN Configuration and verification tasks

Define required VPNs
Configure BGP PE to PE Routing Sessions
Configure BGP PE to CE Routing Sessions
Configure RIP PE to CE Routing Sessions
Configure Static Route PE to CE Routing Sessions
Verify VPN Operations

various pieces of the puzzle to insure we have an appreciation of the manner by which different devices interact with one another.

Figure 4.12 illustrates the flow of traffic between two customer locations via a service provider network. The flow of traffic occurs over a virtual private network (VPN) created via MPLS labels.

In examining Figure 4.12 note that the customer edge (CE) router represents a router at a customer premise that connects to a service provider via a provider edge (PE) router. The provider core (PC) routers interconnect PE and other PC routers and because they support MPLS can be considered to represent label switch routers. The dashed line represents the VPN to be established over the service provider network that connects two customer networks together by providing a virtual path between the CE routers.

4.4.3 MPLS-based VPN attributes

In a Cisco environment each site that belongs to a VPN is distinguished by a number. That number is configured as an 8-byte router distinguisher (RD), which is used to prefix the IP address of the site. The RD is configured on the interface and results in a 12-byte routing prefix.

To transport 12-byte prefixes requires the use of the multi-protocol extension to BGP4, which is MBGP. MBGP must be operational as its function is to propagate information about the VPN to other routers that have interfaces with the same RD value.

When a packet arrives at the PE router it looks up the VPN destination prefix in the MBGP routing table to ascertain the next hop router, which is the exit PE

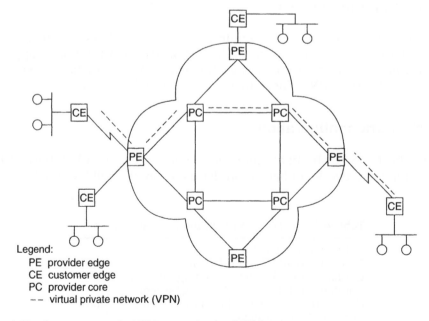

Figure 4.12 Components of a VPN created using MPLS

router. This is because entry PE routers need to deliver packets to the applicable exit PE router and the use of MBGP provides the destination. Although normal IP routing would use the Interior Gateway Protocol (IGP) to determine how to get to the next router, because we are using MPLS, we have a label that routes the packet to the MBGP next hop router, which is the exit PE router. This means that the label provides the routing information and the core routers do not have to be concerned with IP addresses nor examine any information in the IP header.

To facilitate the operation of VPNs a per-site or per-interface routing table is used. That table is referred to as the virtual routing/forwarding instance (VRF). Thus, to create a VPN the CE router needs to have a VRF. The VRF is configured by creating it, describing it to the applicable router, and associating it with one or more interfaces as it can be shared by multiple organizational sites that connect to the same PE router.

Defining VPNs

The first task required in the VPN configuration process is to define VPN routing. In doing so you need to perform a sequence of five steps on the VPN provider edge service (PE) router. Those steps include entering the VPN routing/forwarding instance (VRF) name. As previously noted, a VRF defines the VPN membership of an organization's site that is attached to a PE router. Once the VRF is named you would create routing and forwarding tables. Next, you would create a list of import and/or export route target communities for the specified VRF. Optionally, you could associate the specified route map with the VRF. Last but not least, you would associate the VRF with an interface or subinterface. Table 4.8 lists the commands you would use as well as the rationale for each command required to define VPN routing instances on the PE router.

To illustrate an example of the VPN definition process let us assume an ISP has the Autonomous System (AS) number 123. Let us further assume the ISP has the well-known customer x that requires a VPN and the ISP will create a VRF named vrf00001 and associate it with the Route Distinguisher (RD) 123:1. The ISP will import and export routes to the extended community 123:1 which represent other sites that make up customer x's VPN.

Table 4.8 VPN definition steps

Command	Rationale for use
`Router(config)# ip vrf <vrf-name>`	Define the VPN routing instance by assigning a VRF name
`Router(config-vrf)#rd <route-distinguisher>`	Create routing and forwarding tables
`Router(config-vrf)# route-target import|export| both} <route-target-ext-community>`	Create a list of import and/or export route target communities for the specified VFR
`Router(config-vrf)# import map <route-map>`	Optionally associated the specified route map with the VRF
`Router(config-if)# ip vrf forwarding <vrf-name>`	Associate a VRF with an interface or subinterface

To make the configuration process a bit more interesting let us further assume the ISP has a second customer named y. For that customer the ISP will create a VRF named vrf00002 with an RD of 123:2. Although the ISP will import and export the extended community 123:2 to other customer y sites, let us also assume that customer y needs to reach customer x sites as an extranet partner. To do so requires the ISP to important routes from extended community 123:1 according to a route map named vrf00002. To accomplish the preceding, the configuration would be as follows:

```
ip vrf00001
 rd 23:1
 route-target both 123:1
ip vrf00002
 rd 123:2
 route-target both 123:2
 route-target import 123:1
 import map vrf00002-import-map
```

Configuring BGP PE to PE routing sessions

The second step in the five or six step process when verification is included is to configure BGP PE to PE routing sessions. To accomplish this activity you need to perform a series of three steps. First, you need to configure the Interior BGP (IBGP) routing process with the autonomous system number passed along to other IBGP routers. To accomplish this task you would enter the following command:

```
router(config)#router bgp<autonomous-system>
```

Once you configure the IBGP routing process the next step is to specify a neighbor's IP address or IBGP peer group which functions as a mechanism for identifying it to the local autonomous system. To do so you would enter the following command:

```
router(config-router#neighbor [<ip-address>]<peer-
group-name>]remote-as<number>
```

The third and final step required for configuring BGP PE to PE routing sessions is to activate the advertisement of the IPv4 address family. To accomplish this task you would enter the following command:

```
router(config-router)#neighbor<ip-address>activate
```

Configuring BGP PE to CE routing sessions

The next task required to configure and verify MPLS VPNs is to configure BGP PE to CE routing sessions. This task also represents a three-step process that is performed on the PE router. The first step is to configure an exterior BGP (EBGP) routing process with the autonomous system number being passed to other EBGP routers. To accomplish this step you would enter the same command as for configuring the BGP PE to PE routing session. That is, you would enter the following command:

```
router(config)#router bgp<autonomous-system>
```

The second step requires you to specify a neighbor's IP address or EBGP peer group which functions as a mechanism to identify it to the local autonomous system. To accomplish this you would use the following command:

```
router(config-router)#neighbor{<ip-address>1<peer-
group-name>] remote-as<number>
```

The third step required for configuring BGP PE to CE routing sessions is to activate the advertisement of the IPv4 address family. To accomplish this you would use the following command:

```
router(config-router)#neighbor<ip-address>activate
```

Configuring RIP PE to CE routing sessions

The third task required to configure and verify VPNs using MPLS is to configure RIP PE to CE routing sessions. This task also represents a three-step process that is performed on the PE router.

The first step requires you to enable RIP. To do so you would enter the following command:

```
router(config)#router rip
```

The second step involves defining RIP parameters for PE to CE routing sessions. To accomplish this step you would enter the following command:

```
router(config-router)#address-family
pv4 [unicast] vrf<vrf-name>
```

Concerning the use of the prior command, it should be noted that the default is OFF for auto-summary and synchronization in the VRF address-family submode.

The third step in the configuration process is to enable RIP on the PE to CE link. To accomplish this you would enter the following command:

```
router(config-router)#network<prefix>
```

Configuring static route PE to CE routing sessions

The fifth task to be performed requires you to configure static route PE to CE routing sessions on the PE router. This task requires you to use four commands. The first command involves defining static route parameters for every PE to CE session. To do so you would use the following command:

```
router(config)#ip route vrf<vrf-name)
```

Once you define static route parameters for every PE to CE session you need to define static route parameters for every BGP PE to CE routing session. To do so you would again use the address-family command as follows:

```
router(config-router)#address-family ipv4
[unicast]vfr<vfr-name>
```

Similar to configuring RIP PE to CE routing sessions, when configuring static route PE to CE routing sessions the default is OFF for auto-summary and synchronization in the VRF address-family submode.

The third step in the static route PE to CE routing sessions configuration process is to redistribute VRF static routes into the VRF BGP table. To do so you would enter the following command:

```
router(config-router)#redistribute static
```

The fifth and final step in the configuration of static route PE to CE routing sessions requires you to redistribute directly connected networks into the VRF BGP table. To do so you would enter the following command:

```
router(config-router)#redistribute static connected
```

Verifying VPN commands

In concluding our initial discussion of VPN commands required for the use of MPLS it is always a good idea to verify your efforts. In doing so you can use the 'show' command with applicable keywords and/or parameters to view the result of your prior efforts. In concluding this section we will focus our attention upon the use of the 'show' command to verify the VPN configuration.

The show ip vrf command

To display the set of defined VRFs (VPN routing) forwarding instances you would enter the 'show ip vrf command by itself, without any options. This results in the display of information about all configured VRFs.

If you want to display information about a specific VRF and associated interfaces you would enter the 'show ip vrf' command using the following format:

```
show ip vrf [{<brief>1<detail>1<interfaces}] [<vrf-
name>] [<output-modifiers>]
```

In general, the use of the 'show ip vrf' command to obtain information about a specific VRF would result in the use of the command with the vrf-name being specified. The use of the keyword 'brief' results in the display of concise information on the indicated VRF and associated interfaces while the use of the keyword 'detail' provides, as you might expect, detailed information. The use of the optional keyword 'interfaces' displays detailed information about all interfaces bound to a particular VRF.

The show ip route vrf command

To display the routing table for a VRF you would enter the following command:

```
show ip route vrf <vrf-name>
```

The actual 'show ip route vrf' command has a large number of keyword and variable options. For example, you can use the keyword 'profile' to display a profile of the IP routing table while the keyword 'static' displays static routes and the keyword 'summary' displays a summary of routes. The full format of the 'show ip route vrf' command in all its glory is shown below:

```
show ip route vrf <vrf-name>[connected]
[<protocol>[<as-number] [<tag>] [<output-modifiers>]
[list<number>[<output-modifiers>] [profile]
[static[<output-modifiers>]] [summary
[<output-modifiers>]] [supernet-only
[<output-modifiers>]] [traffic-engineering
[<output-modifiers>]]
```

In examining the full format of the 'show ip route vfr' command it should be noted that you can display applicable output modifiers through the use of the question mark (?) parameter. Concerning keywords not previously mentioned, 'connected' displays all connected routes in a VRF while 'list' specifies the IP access list to display. The use of the keyword 'supernets-only' results in the display being limited to supernet entries, while the use of the keyword 'traffic-engineering' results in the display of traffic engineered routes.

The show ip protocols vrf command

To display routing protocol information for a VRF you would use the 'show ip protocols vrf' command whose format is:

```
show ip protocols vrf <vrf-name>
```

This occurs where the `<vrf-name>` parameter represents the name assigned to a VRF.

The show ip cef vrf command

Continuing our examination of the use of 'show' commands, to display the CEF forwarding table associated with a VRF you would use the following version of the show command:

```
show ip cef vrf <vrf-name>
```

Similar to the 'show ip route vrf' command, the 'show ip cef vrf' command has a significant number of options. The full format of the command is shown below:

```
show ip cef vrf <vrf-name>[<ip-prefix>[<mask>
[longer-prefixes]][detail][<output-modifiers>]]
[<interface><interface-number][adjacency [<interface>
<interface-number>][detail][discard][drop]
[glean][null][punt][<output-modifiers>]]detail]
[<output-modifiers>]][non-recursive][detail]
[<output-modifiers>]][summary[<output-modifiers>]
[traffic[prefix-length][<output-modifiers>]][unresolved
[detail[<output-modifiers>]]]
```

In the preceding format the `<ip-prefix>` variable is optional and represents the prefix in dotted decimal format while the `<mask>` variable represents the mask of the IP prefix, again in dotted decimal format. The optional keyword 'long-prefixes' results in the display of table entries for all of the more specific routes while the keyword 'detail' results in the display of detailed information for each CEF table entry.

When the `<interface>` variable is specified, it denotes the type of interface to use. Acceptable values include ATM, Ethernet, Loopback, POS (Packet Over SONET) or Null. In comparison, the `<interface-number>` denotes the network interface to use.

The keywords 'adjacency', 'discard', 'drop', 'glean', 'null', and 'punt' effect the manner by which prefixes resolved through adjacency are displayed. The keyword 'adjacency' results in the display of all prefixes resolving through adjacency. In comparison, the keyword 'discard' discards adjacency, 'glean' gleans adjacency and so on.

The keyword 'non-recursive' results in the display of non-recursive routes while the keyword 'summary' displays a CEF table summary. The keyword 'traffic' results in the display of traffic statistics whereas the use of the keyword 'prefix-length' results in the display of traffic statistics by prefix-length. Finally, the keyword 'unresolved' displays unresolved routes.

The show ip interface command

Another version of the 'show' command you can consider using displays the VRF table associated with an interface. This command is the 'show ip interface' whose format is shown below:

```
show ip interface <interface-number>
```

The show ip bgp vpnv4 command

Two additional show commands you may wish to use display information about BGPs and label forwarding entries that correspond to VRF routes advertised by the router you are using. To display information about all BGPs you would use the following command:

```
show ip bgp vpnv4 all
```

The show tag-switching command

To display label forwarding entries that correspond to VRF routes advertised by the router you are using you would use the following command:

```
show tag-switching forwarding vrf <vrf-name>
[<prefix mask><length>[detail}
```

In the preceding format the use of the keyword 'detail' results in the display of detailed information concerning the VRF routes.

5

THE RESOURCE RESERVATION PROTOCOL

In this chapter we will focus our attention upon one of the first attempts of the Internet community to provide a Quality of Service (QoS) mechanism within a network. That attempt is in the form of the Resource Reservation Protocol (RSVP), a signaling protocol which enables end systems to request a QoS level from the network. As a result of the use of RSVP, hosts can reserve bandwidth, which allows them to obtain a desired service level or for communicators, what is referred to as a quality of service. Although RSVP does not directly control QoS, it provides information in the form of messages that enable each node to decide whether or not to honor the requested service. While RSVP can work quite well when implemented on an organizational intranet, due to problems of scaling as well as the question of billing for reserved bandwidth between Internet Service Providers (ISPs), it is not implemented on an Internet wide basis and will probably never be implemented on that scale. However, on an intranet basis it provides another mechanism for network managers and router administrators to consider in their quest for QoS.

Similar to previous chapters in this book, this chapter on RSVP contains two sections. In the first section we will examine how RSVP operates. This examination will include looking at the types of messages used under RSVP and their flow. In the second section in this chapter we will examine the manner by which RSVP is implemented in a Cisco hardware environment. In doing so we will examine the applicable commands required to enable RSVP on a router interface.

5.1 UNDERSTANDING RSVP

RSVP represents a signaling protocol used by a host to request its specific requirements from a network. In doing so, RSVP requests represent the requirements for a particular application data stream or flow of packets that have some common characteristics, such as the same destination address and destination port number. RSVP is also used by routers which propagate QoS requests to all nodes along the path of the flows as well as establishing and maintaining sufficient resources for the requested service once that service is approved.

It is important to note that as a signaling protocol RSVP does not perform routing. Instead, RSVP works in conjunction with routing protocols and installs the equivalent of dynamic access list entries along the routes that routing protocols compute.

5.1.1 Overview

RSVP represents a receiver-driven signaling protocol. Although a sender may communicate the availability of an audio, video, or another type of flow of interest to the receiver, it is up to the receiver to request applicable resources. Those resources are requested in only one direction. Thus, we can say that RSVP resource requests are for simplex flows.

Receiver driven signaling

We can obtain an appreciation for the rationale for a receiver driven signaling protocol by considering the manner by which a multicast traffic flow can reach a range of different receivers. For example, consider Figure 5.1 that illustrates the potential flow of a multicast session. In this example note that some links do not transfer multicast traffic since there are no participants in the multicast session located on the route through the link.

In this example only stations A, B, D, F, G, and H are assumed to be participants in a multicast session. While initially the availability of the session

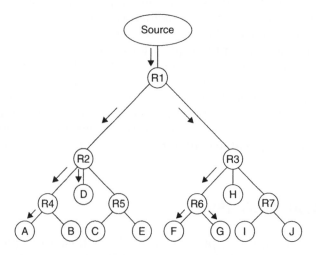

Figure 5.1 An example of multicast routing. In this example stations A, B, D, F, G, and H are assumed to be participants in a multicast

could flow to all stations, when the previously mentioned stations are the only ones to respond positively to join the session, the multicast protocol in each router 'prunes' that portion of the tree structure that does not have any participants as a mechanism to stop the transfer of unnecessary traffic. Thus, in this example the link from routers R2 to R5 would be pruned as well as the link between routers R3 and R7.

Also note that the operating rate of each of the links connecting different pairs of nodes can vary. This means that some members of the multicast group may have different resource requirements than other members of the group. Thus, if multicast traffic can be divided into subflows, some members of a multicast group may require the receipt of only one or a few subflows instead of the cumulative multicast traffic. Similarly, if there are multiple sources that generate traffic into a multicast group, some receivers may only require the ability to receive a portion of the multicast traffic in the form of a subset of resources. Based upon the preceding it is obvious that it makes more sense for the receiver to request the reservation of resources instead of the sender.

RSVP reservation types

In the wonderful world of multicast transmission there are two distinct types of multicast flows. One type of multicast flow originates from exactly one sender and is referred to as a distinct reservation. In comparison, a flow that originates from multiple senders is referred to as a shared reservation.

Distinct reservation

A video application in which each sender transmits a distinct data stream that requires admission and management in a queue represents an example of a distinct reservation. Thus, a video flow will require a separate reservation for each sender on each transmission facility it crosses. Under RSVP this distinct reservation is referred to as an explicit reservation which is installed using a fixed filter style of reservation. The fixed filter (FF) specifies a distinct reservation created for data packets from a particular sender. The reservation scope is defined by an explicit list of senders while the total reservation on a link for a given session is represented by the total of the FF reservations for all requested senders. Here the term scope references the set of sender hosts associated with a given reservation request. If FF reservations are requested by different receivers but select the same sender they must be merged to share a single reservation.

Shared reservation

An example of a shared reservation is an audio application, such as a conference call. Although a conference call results in each sender also emitting a distinct data stream that requires admission and management in a queue, only when

Scope	Reservation Type	
	Distinct	Shared
Explicit	Fixed-Filter (FF) Style	Shared-Explicit (SE) Style
Wildcard	N/A	Wildcard Filter (WF) Style

Figure 5.2 RSVP supports both distinct and shared reservations

persons are not courteous will two or more senders talk (transmit) at the same time. Thus, a shared reservation does not need a separate reservation per sender. Instead, on an as-needed basis, a single reservation is applied to each sender in the set of senders.

Under RSVP a wildcard filter (WF) or shared explicit (SE) style of reservation is supported for shared reservations. The wildcard filter reserves bandwidth and delay characteristics for any sender and is limited by the list of source addresses transported in the reservation message. In comparison, the shared explicit reservation style is used to identify flows for specific network resources. That is, the SE-style reservation creates a single reservation into which flows from all upstream senders are mixed. Similar to a fixed filter style of reservation, the set of senders for an SE style and therefore the scope are specified explicitly by the receiver making the reservation. Figure 5.2 summarizes distinct and shared RSVP reservation style types in the context of their scope.

5.1.2 Traffic support

As a receiver-driven signaling protocol, RSVP receivers request a specific QoS from the network. RSVP operates with both unicast and multicast traffic. You can consider RSVP as a transport protocol as it operates on top of IPv4 or IPv6. However, instead of transporting application data, RSVP transport control messages that inform nodes along a path of the need to allocate resources. Thus, many persons consider RSVP to be similar to the Internet Control Message Protocol (ICMP) or a variety of routing protocols that are also transported by the Internet Protocol (IP). Because RSVP must operate on hosts and routers let us turn our attention to each.

5.1.3 Host operations

Figure 5.3 illustrates the manner by which RSVP operates in a host computer. As indicated in Figure 5.3, the major RSVP modules operating on a host include the application that generates an RSVP request which determines if the request should be honored, admission control which determines if sufficient resources are available, a classifier that categorizes the request, and a packet scheduler which is responsible for allocating packet transmission.

An RSVP compatible application will pass a QoS request to an RSVP process operating on the local host. During this initial process the RSVP protocol will

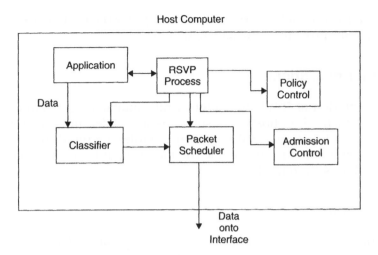

Figure 5.3   RSVP modules operating on a host computer

also transport the request to all of the nodes in the reserve path from the data source. The reason the reverse path is used is due to the fact that as previously noted, RSVP is receiver driven.

Reservation set-up

During the reservation setup process each RSVP QoS request is passed to two decision modules operating on the host. Those modules are admission control and policy control.

Admission and policy control

The function of the admission control module is to determine if the host has sufficient resources to honor the requested QoS. In comparison, policy control is tasked with determining if the user has administrative permission to make the reservation. This means that the LAN administrator must add RSVP permission to the large number of operations they perform. Otherwise, it could be possible for a user to request 784 Kbps of bandwidth to facilitate the downloading of music files, a situation not conducive to network operations. If both a check via admission control and a check via policy control are passed, parameters will be set in the packet classifier and packet scheduler. Otherwise, if either check fails, the RSVP process will return an error indication to the application that originated the request.

Classifier and scheduler

Assuming an RSVP QoS request is passed by admission control and policy control, data flows into the classifier. In the classifier the QoS class is determined for each packet. For each outgoing interface the packet scheduler is responsible for providing the allocated QoS.

5.1.4 Router operations

The implementation of RSVP on routers while similar to the manner by which RSVP is implemented on hosts has some distinct differences. Figure 5.4 illustrates the major functional RSVP related modules required for this signaling protocol to operate on a router. While the major modules remain the same, the flow of data differs from the host.

Data flow actions

As RSVP messages transporting reservation requests originate at receivers, they are transmitted upstream towards the sender. At each intermediate node a reservation request results in the occurrence of two actions. First, the RSVP process operating on the node passes the request to admission control and policy control modules. If either module rejects the request the reservation request will be rejected. If the reservation request is rejected, the node will return an error message to the receiver. If the reservation request is honored by both admission control and policy control the packet classifier is then instructed to select data packets defined by what is referred to as a flowspec. Thus, prior to continuing our discussion of the reservation process, a brief digression concerning the flowspec and a related topic referred to as a filterspec is in order.

The flowspec

Under RSVP operations are based by handling a sequence of interrelated packets as an entity. A set of packets with a common characteristic, such as

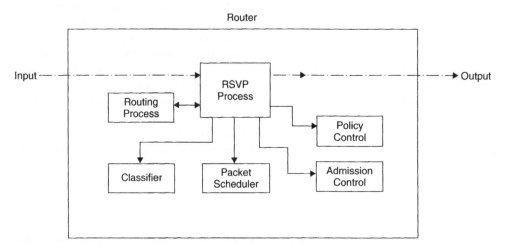

Figure 5.4 RSVP router modules

the same destination address and destination port number, is treated as an entity and referred to as a flow. Under RSVP the treatment of the flow is referred to as a flow specification or simply a flowspec. That is, the flowspec describes the traffic transmitted as well as the service requirements of an application, which results in a request for a desired QoS. The term 'desired QoS' is used since the flowspec does not have to be honored. Thus, a flowspec can be considered to represent a reservation request.

The flowspec in a reservation class will normally include a service class and two sets of numeric parameters. The service class is defined by the application. The first parameter defines the desired QoS and is referred to as an Rspec, with 'R' for 'reserve'. The second parameter describes the data flow. This parameter is referred to as a Tspec, with 'T' denoting 'traffic'. The Tspec can be considered to represent one side of a 'contract' between the data flow and the service. That is, once a router accepts a specified QoS, it must continue to provide that level of service as long as the flow is within the boundary of the Tspec. However, if traffic should exceed the expected level, the router can then drop packets, revert to servicing the flow on a best-effort basis, or employ another service mechanism.

The filterspec

The actual reservation request consists of a flowspec and a filterspec. As indicated previously, the flowspec denotes the desired QoS. In comparison, the filterspec specifies those packets that will be serviced by the flowspec. Thus, the filterspec defines the specific type of traffic that will be provided with a QoS. In comparison, the flowspec defines the QoS class of service to be provided.

Figure 5.5 illustrates the general relationship between the filterspec and the flowspec as data flows through an RSVP compatible router.

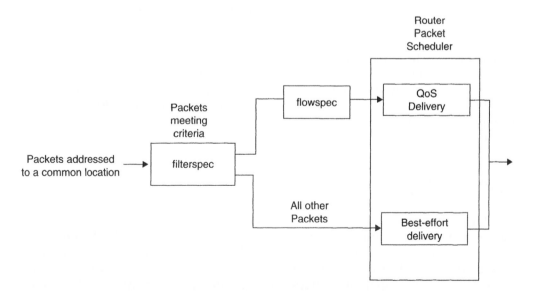

Figure 5.5 The general relationship between the filterspec and flowspec

In examining Figure 5.5 note that the filterspec operates on a flow prior to the flowspec operating upon any packets. The filterspec subdivides packets based upon matching them against a predefined criteria. Packets that meet the predefined criteria will require QoS delivery. Thus, such packets are passed to the flowspec to determine the manner by which QoS delivery is to be provided. In comparison, packets that do not meet the predefined criteria are passed to the router's packet scheduler for best-effort delivery.

Now that we have an appreciation of the manner by which the flowspec and filterspec govern the reservation on a link, let us refocus our attention upon the reservation process. In doing so let us turn our attention upon the second action a reservation request triggers. That action is, when applicable, for a node to forward the request upstream.

Forwarding the reservation request

The second action triggered by a reservation request is the forwarding of the request in the upstream direction. That request is propagated to the appropriate senders, with the set of sender hosts associated with a given reservation request referred to as a scope.

As a reservation request is forwarded upstream the traffic control mechanism at each node may modify the flowspec on a hop-by-hop basis. In addition, each node examines reservations from downstream branches of the same multicast tree from the same sender or set of senders and merges multiple reservations as reservations travel upstream. When a receiver originates a reservation it can optionally request a confirmation that the request was honored by the network. A successful reservation request will propagate in the upstream direction along the multicast tree until it reaches a location where an existing reservation is equal or greater than that being requested. At that location the arriving request will be merged with the reservation in place and does not need to be forwarded. The node at that location can send a reservation confirmation message back to the receiver.

While we like to consider a confirmation as a guarantee, an RSVP confirmation is similar to one for an airline flight. That is, airlines periodically overbook and when too many passengers show up a problem occurs, although such a problem is infrequent unless someone is sitting in your seat. Similarly, the receipt of an RSVP confirmation represents a high-probability indication and not a guarantee.

RSVP tunneling

Similar to the deployment of any new protocol, the deployment of RSVP cannot be accomplished instantaneously. Instead, RSVP may be periodically deployed on an intranet and could even be deployed on small portions of the Internet. Due to this fact, RSVP must provide correct protocol operations even in the event two RSVP compatible routers are separated from one another by one or more noncompatible RSVP routers. Although the noncompatible RSVP routers are obviously unable to provide resource reservations, if the transmission links through those routers have a sufficient amount of extra capacity, it is highly possible to route packets through non-RSVP compliant routers and still obtain a desired level of service.

RSVP supports connectivity via non-RSVP routers through the process of tunneling. In the tunneling process both RSVP and non-RSVP routers forward RSVP path messages toward their destination address via the use of local routing tables. Here the path message represents a message transmitted by a sender towards a receiver. In comparison, each receiver that wants to participate will return a reservation request message in the opposite direction. When a path message traverses a group of non-RSVP routers representing a non-RSVP cloud, the path message will copy the IP address of the last RSVP-compatible router. The other primary type of RSVP message is the reservation message. When encountering a non-RSVP cloud, reservation requests simply flow through the cloud to the next upstream RSVP-compatible router. RSVP tunneling is transparently performed. This means that an organization placing RSVP into effect on two intranets connected by an Internet VPN transmission facility may or may not receive an end-to-end QoS based upon the level of utilization of the Internet VPN transmission facility.

5.1.5 Message types and formats

A generalized information format is used by RSVP to transfer information between RSVP compliant nodes. Each RSVP message consists of a common header followed by one or more object records. As we will shortly note, a field within the header defines the message type, while the object records transport specific information relevant to the message.

The RSVP header

Figure 5.6 illustrates the RSVP header that contains information fields that describe the actions or events defined by an RSVP message.

Version and flags fields

The version field is 4 bits in length and defines the current version of the RSVP protocol. The version field is followed by a 4-bit flags field. This field functions as a reserved placeholder that will allow new features to be added to the protocol at a later date.

Message type field

The message type field is 8-bits in length. The value of this field defines the type of message being transmitted. The type of messages currently supported

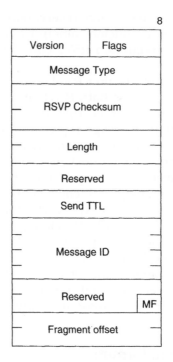

Figure 5.6 RSVP message header fields

are indicated below:

> 1 = Path
>
> 2 = Reservation-request
>
> 3 = Path-error
>
> 4 = Reservation-request error
>
> 5 = Path-teardown
>
> 6 = Reservation-teardown
>
> 7 = Reservation-confirmation

Based upon the above list of currently defined message types we can subdivide RSVP messages into two categories – path management and reservation management. Path management messages (values 1, 3 and 5) are originated by the source of the real-time transmission in the downlink direction. Such messages permit potential recipients as well as downstream routers to gather information on the traffic route to the source. Reservation messages (types 2, 4, 6 and 7) are used by the recipients and downstream routers to negotiate a reservation. The relationship between the path and reservation messages can be considered to represent one of an inverted IP address. That is, the reservation message is sent by each receiver in the opposite direction toward the sender by reversing the paths of the path messages.

Teardown messages

RSVP 'teardown' (Path-teardown (PathTear) and Reservation teardown (ResvTear)) messages are used to remove the path or reservation state. The PathTear message travels towards all receivers downstream from its point of initiation and deletes path state and all dependent reservation state on its flow. The ResvTear message flows towards all senders in the upstream direction from its point of initiation, deleting reservation state. A teardown request can be initiated by an application in the sender or receiver or by a router due to a timeout condition. Once initiated, the teardown request is forwarded on a hop-by-hop basis. Because teardown requests are not delivered reliably, a timeout value is also used by routers. That is, if no RSVP message is received within a predefined period of time the router will initiate a new teardown message.

Error messages

Similar to teardown messages there are two types of error messages – Reservation-error (ResvErr) and Path-error (PathErr). PathErr messages are transmitted upstream to the sender that created the error and do not change the path state in the nodes they traverse. In comparison, a ResvErr message results from a request that fails admission control.

Checksum field

The RSVP Checksum field is 16-bits in length. This checksum represents a standard TCP/UDP checksum that is computed over the contents of the RSVP message.

The function of this field is to ensure the correct delivery of information. Note that this field is not a CRC but a checksum, with the value of the checksum is computed as the one's complement of the one's complement sum of the message. When RSVP is transmitted over an unreliable transport service, the checksum field can be set to a value of all-zeros to indicate that no checksum was transmitted.

Length field

The RSVP length field is 16-bits in length. This field indicates the total length of the RSVP message in bytes. That length includes the common message header as well as the variable-length object records that follow. If the More Fragment (MF) one-bit flag is set in the header or the fragment offset field value is non-zero, then the value of the length field represents the length of the current fragment of a larger message.

Send TTL field

The Send TTL field is 8-bits in length. This field contains the IP header Time to Live (TTL) value with which the RSVP message was sent and can be used if a message was processed by non-RSVP capable nodes.

Object records

One or more variable length object records follows the RSVP header. The purpose of the object record is to convey information about the session, the IP address of the node that sent the message, time values at which the originator will update the message, and flowspec and filterspec information. In addition, an object record can also contain the scope for reservation, reservation-error, or reservation-teardown messages and a reservation confirmation in a reservation or reservation confirmation message. Each of the seven basic RSVP message types will have its header followed by one or more object records to fully define the message.

Object formats

Each object consists of one or more 32-bit words (4 bytes) with a one-word header. Figure 5.7 illustrates the format of an object record.

Length field

The 16-bit length field defines the length of the object record in bytes. The value of this field is always a multiple of 4 and have a value of at least 4.

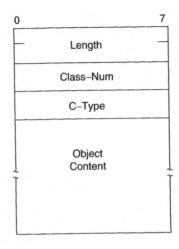

Figure 5.7 The format of an object record

Table 5.1 Object class identifiers

Type	Description
NULL	Ignored by receiver.
SESSION	Contains IP destination address, IP protocol identification and destination port.
RSVP_HOP	Transports the IP address of the RSVP-capable node that sent the message.
TIME_VALUES	Contains the refresh period value for each Path and Resv message.
STYLE	Defines information not in flowspec or filterspec.
FLOWSPEC	Defines a desired QoS in a Resv message.
FILTER_SPEC	Defines packets that should receive the desired QoS.
SENDER_TEMPLATE	Contains the sender IP address to identify the device.
SENDER_TSPEC	Defines the traffic characteristics of the sender's data flow.
ADSPEC	Transports advertising data in a path message.
ERROR_SPEC	Specifies an error in a PathErr, ResvErr or a ResvConf message.
POLICY_Data	Transports information that enables a local policy nodule to determine if an associated reservation is permitted.
INTEGRITY	Transports cryptographic data to authenticate the originating node and verify the contents of the RSVP message.
SCOPE	Transports an explicit list of sender hosts towards which information in the message is to be forwarded.
RESV_CONFIRM	Transports the IP address of a receiver that requested a confirmation.

Class-Num field

The Class-Num field identifies the object class. There are 15 distinct types of object classes with each class having a distinct name. Those object classes are listed in Table 5.1.

C-Type field

The function of the C-type field is to define the object type uniquely within the Class-Num field. The Class-Num and C-Type fields can be used together as a 16-bit number to define a unique type for each object. Because the primary purpose of this section is to obtain a general understanding of the manner by which RSVP operates, we will not probe deeper into the object fields. Instead, we will now turn our attention to examining how RSVP is configured in a Cisco environment.

5.2 CONFIGURING RSVP

In the first part in this chapter we became acquainted with the manner by which RSVP operates. Now that we have a basic understanding of RSVP we will put theory into practice by examining applicable RSVP configuration commands in a Cisco environment. As we will shortly note in this section, although RSVP is a mildly complex protocol, when an applicable planning process occurs, at the most the configuration process results in the issuance of a small number of commands that only support a few optional parameters. Because the use of RSVP may not be suitable for different types of transmission facilities we will first focus our attention upon its suitability for different transmission links. Once this is accomplished we will note the tasks required to implement RSVP as well as review the use of the applicable commands to implement RSVP in a Cisco environment.

5.2.1 RSVP link suitability

When considering the use of RSVP it is important to note that this protocol requires the use of a transmission facility that supports queuing on nodes. For this reason serial line interfaces to include frame relay and ATM connections as well as the use of such protocols as Higher Level Data Link Control (HDLC) and Point-to-Point Protocol (PPP) are well modeled by RSVP. In comparison, the ability to use RSVP on a public X.25 network or a multiaccess LAN is problematic as no useful reservation can be made in either type of network environment.

Frame relay considerations

When implementing RSVP on a frame relay network your reservations are made on an interface or subinterface basis. It is important to note that a subinterface can contain more than one Data Link Circuit Identifier (DLCI). Here the DLCI results in the creation of a virtual path from the interface or subinterface through the frame relay network. Each interface or subinterface can support multiple DLCIs, in effect providing multiple virtual circuits into and through the frame relay network. If so, the required bandwidth and the reserved bandwidth could differ. Due to this situation RSVP subinterfaces of frame relay circuits must be configured for only one DLCI to operate correctly. In addition, because of the manner by which frame relay DLCIs operate, it is important to note that the Committed Information Rate (CIR) and Committed Burst size (Bc) and Excessive Burst (Be) may not be reflected in the configuration and, if reflected, can differ from the interface speed. This means that when you use the 'ip rsvp bandwidth' command that we will examine later in this section, you need to use this interface configuration command for both the interface and the subinterface.

As a quick frame relay review, the physical interface speed defines the maximum permissible data rate into a frame relay network. For example, a T1 circuit operates at 1.544 Mbps that represents the maximum operating rate of the circuit. In actuality, a T1 line has a 8 Kbps framing signal. Thus, the actual data transfer rate on this transmission facility is 1.544 Mbps less 8 Kbps or 1.536 Mbps.

In comparison, the CIR denotes the data rate that the network operator is committed to support. For example, the CIR on a T1 connection could be 128 Kbps, 256 Kbps or even 1.536 Mbps. Because LAN traffic is bursty, only periodically will transmission exceed the CIR. When it does, frames will be marked as eligible for discard, although the probability of a frame being discarded is quite low. When the data rate exceeds the CIR, the sum of the CIR and Be + Bc represents the line operating rate. Thus, there is a physical limit concerning the amount of data that can be burst above the Committed Information Rate.

When using a circuit with multiple DLCIs you must configure the amount of the total interface and the amount of each receiving interface that requires reservation. For example, assume your organization has a T1 connection at corporate headquarters that provides connectivity into a frame relay network. Further assume that the T1 line supports several DLCIs to remote regional offices that are served via 128 Kbps fractional T1 and 56 Kbps digital circuits. When you use the 'ip rsvp bandwidth' command, as we will shortly note, the default maximum bandwidth is 75 percent of the interface bandwidth. This means that you would need to configure 1.158 Mbps which represents 75 percent of the T1 line and 96 Kbps and 42 Kbps, which represent 75 percent of 128 Kbps and 42 Kbps if you want to match the default characteristics of Cisco's RSVP implementation.

ATM considerations

If you are designing RSVP for an ATM internetwork you need to consider a special design consideration. This is because many ATM implementations more than likely use a usable bit rate (UBR) or an available bit rate (ABR) virtual channel (VC) for connecting individual routers. If the UBR or ABR VC is used, ATM makes a best effort to satisfy the bit-rate requirements of traffic and assumes that higher layers operating on the end stations assume responsibility for data that does not get through the network. Both types of ATM services support the ability to open separate channels for reserved traffic having the necessary characteristics. Thus, RSVP needs to open those VCs and adjust cache to make effective use of virtual channels.

5.2.2 RSVP tasks

Once you appropriately plan your RSVP configuration you are ready to configure this protocol on your routers. In doing so you must first configure RSVP on applicable interfaces to use the protocol. Once this step is accomplished you can consider performing one or more of five optional tasks. Table 5.2 lists all six tasks you can consider when developing an RSVP configuration plan. Note that only the first step in Table 5.2 is mandatory.

Now that we have an appreciation of the steps involved in developing an RSVP configuration plan, let us turn our attention to the commands we need to use for the configuration process. In doing so we will follow the order of the six steps previously listed in Table 5.2.

Table 5.2 RSVP configuration plan steps

```
Enable RSVP on an interface.
Enter Senders in the RSVP database.
Enter Receivers in the RSVP database.
Enter Multicast Addresses.
Control which RSVP Neighbor can offer a resolution.
Monitor RSVP.
```

Enabling RSVP

RSVP is disabled by default. To enable RSVP on an interface you need to use the 'ip rsvp bandwidth' command. The format of this global configuration command is shown below:

ip rsvp bandwidth [<interface-kbps>] [single-flow-kbps>]

As indicated by the two optional parameters, this command can be used to set the bandwidth and single-flow limits. The default maximum bandwidth is 75 percent of the interface bandwidth, while a single flow can reserve up to the entire reservable bandwidth. On subinterfaces, this command uses the more restrictive of the available bandwidths of the physical interface and the subinterface. For example, if we return our attention to frame relay, let us consider a T1 interface. That interface operates at 1.544 Mbps, however, 8 Kbps is used for framing which leaves a capacity of 1.536 Mbps. If the DLCIs operate at 128 Kbps, then 75 percent of 128 Kbps would become 96 Kbps.

According to Cisco literature, reservations on individual circuits that do not exceed a data rate of 100 Kbps normally succeed. In comparison, for reservations made on other circuits that add up to 1.2 Mbps, and if a reservation is made on a subinterface that has enough remaining bandwidth, the reservation will be refused as the physical interface in this situation would lack supporting bandwidth. This information from Cisco documentation is questionable and appears to be a typo, since the difference between 75 percent of 1.536 Mbps and 1.2 Mbps is −.05 Mbps.

Enter senders in the RSVP database

Entering senders in the RSVP database represents the first of five optional steps you can consider. If you enter senders in the RSVP database the effect is to configure the router to behave as if it were periodically receiving an RSVP PATH message from the sender or previous hop routes containing certain attributes. To enter senders in the RSVP database you would use the 'ip rsvp sender' command whose format is shown below:

ip rsvp sender <session-ip-address><sender-ip-address>
[tcp|udp|<ip-protocol>] <session-destport><sender-

```
sendport><previous-hop-ip-address><previous-hop-
interface.
```

Enter receivers in the RSVP database

Another optional step is to enter receivers in the RSVP database. Doing so permits you to configure the router to behave as though it is continuously receiving an RSVP RESV message from an originator with one or more indicated attributes. To enter receivers in the RSVP database you would use the 'ip rsvp reservation' command whose format is indicated below:

ip rsvp reservation <session-ip-address><sender-ip-
address>[**tcp|udp**|<ip-protocol>] <session-destport>
<sender-sourceport><next-hop-ip-address><next-hop-
interface[**ff|se|wf**] [**rate|load**] [<bandwidth>] [<burst-
size>]

Here [**ff|se|wf**] references the reservation style. Fixed Filter (FF) is a single reservation. Shared Explicit (se) represents a shared reservation, limited scope, and Wild Card (WC) is shared reservation, unlimited scope. The keywords [**rate|load**] refer to QoS–guaranteed bit rate service or controlled load service.

Enter multicast address

Under RSVP control messages are intended to use raw IP packets, however, some hosts may not be able to transmit such packets. If RSVP neighbors are noted to be using UDP encapsulation, the router will automatically generate UDP-encapsulated messages for its neighbors. However, some hosts will not originate such messages unless they hear from the router first. In this situation you would want to use the 'ip rsvp udp-multicast' command to enter a UDP multicast address. The format of this global command is shown below:

ip rsvp udp-multicast [<multicast-address>]

Control which RSVP neighbor can offer a reservation

By default, any RSVP neighbor can offer a reservation. To limit which routers can offer a reservation you can use the 'ip rsvp neighbors' command whose format is:

ip rsvp neighbors access-list number

When this command is enabled, only neighbors conforming to the access list will be accepted.

Monitor RSVP

The sixth step and fifth optional step in the RSVP configuration process is to monitor your network resource. There are six 'show ip rsvp' commands you can consider using and which we will now examine.

Display RSVP-related interface information

To display RSVP-related interface information you would use the 'show ip rsvp interface' command. The format of this command is:

```
show ip rsvp interface [<interface-type><interface-
number>]
```

Here the <interface-type> represents the name of the interface while the <interface-number> represents the number of the interface.

Display RSVP-related filters and bandwidth

You can display RSVP-related filters and bandwidth information by using the 'show ip rsvp installed' command. The format of this command is:

```
show ip rsvp installed [<interface-type><interface-
number>]
```

This occurs where the two optional parameters were previously described in our discussion of the 'show ip rsvp interface' command.

Display current RSVP neighbors

To display RSVP neighbors you would use the 'show ip rsvp neighbor' command. The format of this command is shown below:

```
show ip rsvp neighbor [<interface-type><interface-
number>]
```

This occurs where the optional parameters were previously described when the 'show ip rsvp interface' command was reviewed.

Display RSVP sender information

You can display RSVP PATH related sender information in the router's database. To do so you would use the 'show ip rsvp send' command whose format is:

```
show ip rsvp sender [<interface-type><interface-
number>]
```

Once again, the optional parameters for this command were previously described when the 'show ip rsvp interface' command was described.

Display RSVP request information

To display RSVP-related request information being transmitted upstream from the router you would use the 'show ip rsvp request' command. The format of this command is:

```
show ip rsvp request [<interface-type><interface-
number>]
```

Display RSVP receiver information

To display RSVP-related received information in the router's database you would use the 'show ip rsvp reservation' command. The format of this command is:

```
show ip rsvp reservation [<interface-type><interface-
number>]
```

Related RSVP command

One RSVP-related command you should consider using is the 'show queue' command. The use of this command with an applicable interface will provide you with statistics concerning weighted fair queuing to include the number of drops. The latter could be used to determine if your current configuration should be modified.

Configuration example

In closing our discussion of RSVP let us turn our attention to the use of various RSVP commands with a sample configuration. Figure 5.8 indicates a sample

```
interface serial s0
  ip rsvp bandwidth 15360 100
  !
  ip rsvp udp-multicast 10.11.12.0
  !
  ip rsvp neighbors 1
  !
  access-list 1 permit 198.78.46.11
  !
```

Figure 5.8 RSVP configuration example

RSVP configuration. In examining the sample RSVP configuration shown in Figure 5.8 it should be noted that this is not a complete configuration and is only for illustrative purposes to show the use of several RSVP commands. Thus, what should be conspicuous by its absence is the assignment of an IP address to the router's sØ interface as well as other common configuration commands. In examining the RSVP related commands listed in Figure 5.8, the first command sets the interface speed to 1.536 Mbps, while the single-flow rate is set to 1 Mbps. The second RSVP-related statement enters a UDP multicast address to ensure that the host at that address will accept UDP-encapsulated messages. The third RSVP-related message limits routers that can offer an RSVP reservation. In this example the 'ip rsvp neighbors' command references access-list number 1. One statement is shown for that access list, which permits IP address 198.78.46.11 which identifies the first of possibly a long list of routers that can offer a reservation.

6

QoS ENHANCEMENT TECHNIQUES

Until this chapter our primary focus of attention was upon industry standards developed specifically to provide a Quality of Service (QoS) or traffic expediting capability through a network. As such, we previously examined the use of the Type of Service (ToS) byte in the IPv4 header and its modern counterpart referred to as Differentiated Service (DiffServ), the use of the dynamic duo of standards used to expedite traffic through Layer-2 switches known as the IEEE 802.1p and 802.1Q standards, network layer switching through the use of Multi-Protocol Layer Switching (MPLS), and the reservation of bandwidth via the use of the RSVP signaling protocol. While each of the previously mentioned QoS or traffic expediting techniques can be of considerable importance in minimizing latency and jitter of packets flowing through a network, by themselves they are only one part, although an extremely important part, of what this author commonly refers to as the 'QoS puzzle'. The other key part of the QoS puzzle can be considered to represent techniques that enhances the flow of data through a network. Thus, the purpose of this chapter is to make readers aware of a variety of techniques you can consider that will enhance the flow of data into and through a network.

Some of the enhancement techniques we will discuss in this chapter include the use of static routing, the configuration of RTP header compression when transporting digitized voice, altering the time TCP waits for a connection prior to timing out and the use of selective acknowledgments. While some of the techniques covered in this chapter may not be applicable to your current operational environment, it is possible that other techniques are. In addition, by knowing about these techniques you may obtain the opportunity to use them at a later date as a mechanism to enhance the flow of data into and through your organization's network.

6.1 ENABLING STATIC ROUTING

In this section we will focus our attention upon the use of static routing as a mechanism to eliminate the transfer of routing table updates. This elimination will reduce a potential impediment to the transfer of information that in turn expedites the flow of data between networks.

6.1.1 Overview

In a large network it can be considered as a given that the network manager will configure one or more routing protocols to automatically facilitate the updating of routes if a communications link should fail. One example of a relatively old but still commonly used routing protocol is the Routing Information Protocol (RIP). RIP represents an Interior Gateway Protocol (IGP) developed for use in small, homogeneous networks.

Routing table updates

RIP represents one of several distance-vector routing protocols available for use. Because the RIP hop count is limited to a maximum of 15, its use is restricted to small networks. RIP employs the User Datagram Protocol (UDP) to exchange routing information among routing neighbors. Each router in a RIP network will transmit its routing table every 30 seconds. The process of transmitting routing table updates is referred to as advertising and can place a fairly large performance burden on low speed communications links. In actuality, the performance burden depends both on the size of the router's routing table and the transmission speed of the communications link over which the contents of the table are advertised. Because the quantity of data in the advertised routing table depends upon the design of the network infrastructure, we can also say that as the complexity of the network increases and more interfaces are added to a router, the size of their routing table increases. Thus, the complexity of a network as well as the number of router interfaces will affect the quantity of data transmitted every 30 seconds under RIP.

Under RIP a directly connected network has a hop count of zero. In comparison, an unreachable network has a hop count of 16, which limits the maximum number of hops supported to 15. If a router does not receive an updated routing table after a period of 180 seconds, it will mark the routes served by the non-updating router as being unusable. If, after a period of 240 seconds (4 minutes), an update has yet to be received, the router will remove all routing table entries associated with the non-updating router. Due to these time delays the failure of a link in a mesh-structured network can result in a relatively long time for a new route to be learned, a process referred to as convergence.

Rationale for eliminating table updates

In spite of generating extra traffic in the form of routing table updates and a relatively slow convergence process, RIP is a commonly used protocol. However, when a private network is connected to the Internet via the services of an Internet Service Provider (ISP), the use of a routing protocol is more than likely unnecessary. Instead, you should configure a static route between your organization's gateway router to the Internet and your Internet Service Provider's gateway.

6.1.2 Working with static routes

As a brief reminder for some readers, static routes represent user-defined routes that cause packets moving between a source and a destination to take a specified path. Because the path is fixed, there is no need to perform routing table updates. However, it is quite possible to either intentionally or inadvertently override a static route by information from a dynamic routing protocol. To do so you need to consider the assignment of administrative distance values. By default, a static route has an administrative distance value of 1. In comparison, routing protocols normally have much higher default administrative distance values. To override a static route by information from a dynamic routing protocol requires an administrative distance value of the static route to be higher than that of a dynamic protocol. Table 6.1 lists administrative distance values for nine route sources.

In examining the default administrative distances listed in Table 6.1 note that other than a connected interface, a static route has the lowest default distance. Thus, to override the assignment of a static route you would need to change its default distance to a value greater than the dynamic routing protocol you wish to use.

Establishing a static route

The command you would use to establish a static route is the 'ip route' command. The format of this command is shown below:

ip route [<mark>]{<address> <interface>}[<distance>]

Table 6.1 Default administrative distance values

Route source	Default distance
Connected interface	0
Static route	1
External BGP	20
IGRP	100
OSPF	110
IS-IS	115
RIP	120
EGP	140
Internal BGP	200

This occurs where the parameter <distance> represents the administrative distance value. The default value for static routing is 1, which results in a static route not being overridden by information from a dynamic routing protocol.

To illustrate the configuration of static routing we need a configuration so let us create one. Figure 6.1 illustrates a small internetwork consisting of a regional office connected to an Internet Service Provider.

As indicated in Figure 6.1, we will assume that the network address of the regional office is 198.78.46.0 while the serial port of the router has the address 205.131.175.1. At the ISP location we will assume that the serial port of the router has the IP address of 205.131.175.2 while the Ethernet network at that location has the IP address of 205.131.155.0.

To set up a static route from the regional office the ISP you would enter the following command:

```
ip route 205.131.155.0 255.255.255.0 205.131.175.2
```

This command tells the router at the regional office to construct a static routing table such that network 205.131.155.0 which represents the LAN at the ISP is reached by the serial connection between the two routers, with the serial interface on the router configured with the IP address 205.131.175.2. At the router at the ISP the 'ip route' command would be required to configure its static routing table in the reserve direction. Thus, when employed on a limited basis where there is only one route between source and destination, the use of static routing is relatively easy to implement and eliminates routing table updates.

Avoiding an inadvertent override

One important point to note concerning the use of static routes is the fact that such routes that point to an interface will be advertised by the use of any dynamic routing protocol, such as RIP and IGRP. This results from the fact that static routes that point to an interface are considered in the routing table

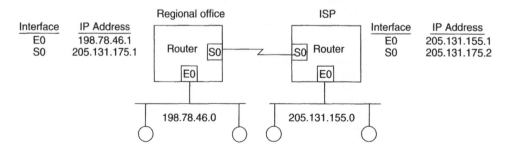

Figure 6.1 An example of a direct connection to an ISP that will be configured for static IP routing

```
Router#config t
Enter configuration commands, one per line. End with CNTL/Z
Router(config)#router rip
Router(config-router)#network 205.131.175.0
Router(config-router)# network 205.131.185.0
Router(config-router)#end
```

Figure 6.2   Configuring RIP for a router connected to two networks

as connections, which results in their loss of a static state. However, if you define a static route to an interface that is not one of the networks defined in a 'network' command when you configure a routing protocol, no dynamic routing protocols will advertise the route. As a refresher, when configuring a routing protocol you use the 'network' command to define the major network numbers connected to a router. For example, assume you are configuring RIP and your router is connected to networks 205.131.175.0 and 205.131.185.0. To configure RIP as well as specify the IP networks connected to the router, you would enter the commands shown in Figure 6.2.

6.2 ENHANCING THE ADDRESS RESOLUTION PROCESS

Our second enhancement technique to be covered in this chapter will permit us to minimize network delays associated with the address resolution process. To ensure we obtain an appreciation of how we can minimize periodic delays associated with the address resolution process, we will first discuss why it occurs prior to noting how we can avoid its delay.

6.2.1 Overview

Although network users surfing the Internet are primarily concerned with obtaining an IP address for their workstation, that address only represents one half of the address story. Usually hidden literally behind the scene is a second address associated with each workstation and router connected to a LAN. That address represents the burnt-in address of the LAN network interface card or a network ROM chip mounted on the motherboard of many computers that performs the network interface functions. Referred to as the Media Access Control (MAC) address, this address represents a Layer 2 address in the Open System Interconnection (ISO) Reference Model. In comparison, the IP address represents a network address or a Layer 3 address with respect to the ISO Reference Model.

Address structures

In addition to the MAC IP addresses representing different layers of the International Standards Organization (ISO) OSI Reference Model, they also represent different physical address structures. The MAC address is a 48-bit address that is divided into two parts. The first 24 bits or 3 bytes of the MAC address represents a manufacturer identification (manufacturer-ID). The manufacturer-ID is assigned by the Institute of Electrical and Electronics Engineers (IEEE), allowing vendors to use the remaining three bytes of the address to uniquely define each network adapter they manufacture. If a manufacturer is successful in fabricating and selling their network adapters, prior to running out of numbers of use in the three-byte position that uniquely identifies each adapter, the manufacturer would apply to the IEEE for an additional manufacturer-ID. The top portion of Figure 6.3 illustrates the format of the 6 byte source and destination address fields in an Ethernet Layer 2 frame used to transport data on an Ethernet LAN.

The first two bit positions represent subfields only applicable to the destination address in an Ethernet frame and are set to a bit value of zero in the source address field. Thus, for an Ethernet network adapter the manufacturer-ID portion of the adapter address will actually be 22 bits in length, since the first two bits are used to define the type of address and how the address is to be interpreted. That is, the setting of the first bit in the destination address field indicates if the address represents an individual address or a group address. The setting of the second bit position denotes whether the address in the field represents a universally administrated address or a locally administrated address.

A universally administrated address represents the address burnt into read-only memory (ROM) on the network adapter. This address is unique globally since the manufacturer-ID portion of the address is, as previously mentioned, controlled by the IEEE. In comparison, a locally administrated address represents an address configured via the use of software when a workstation is powered on. The locally administrated address overrides the burnt-in universally administrated address and allows network managers and LAN administrators to predefine address without having to acquire a large number of network adapters.

The use of locally administrated addressing dates to the period when it was a time-consuming process to configure communications controllers attached to mainframes to recognize new workstations. The communications controller would have to be literally taken down to be reconfigured, potentially resulting in

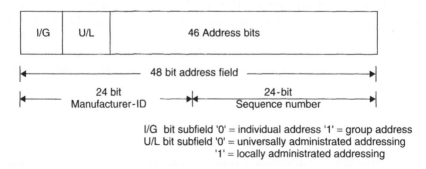

Figure 6.3 Ethernet source and destination address field formats

hundreds to thousands of terminals losing their connectivity to the mainframe. By using locally administrated addresses it became possible to pre-configure groups or blocks of addresses for use without having to purchase network adapter cards prior to the cards being needed. This in turn enables the network manager to add workstations to a mainframe-based network without having to bring down the communications controller and the network the controller supports. Although the use of locally administrated addresses was popular during the 1980s, its use has considerably diminished since the communications controller can now be reconfigured without having to bring down the attached network.

LAN frame structure

On a LAN data is delivered in the form of frames. Such frames use the Layer-2 MAC address for data delivery. For example, consider the format of an Ethernet frame, which is shown in Figure 6.4.

The 48-bit destination address field represents the MAC address of a workstation on a LAN where a frame will be delivered, while the source address field represents the MAC address of the originator of the frame.

IP address structure

In comparison to the MAC address that is used for data delivery on LANs, the Internet Protocol (IP) uses a 32-bit address for the routing of data through an IP network. There are five classes of IP addresses, referred to as Class A through Class E, with Classes A, B, and C subdivided into network and host sub-addresses. A Class A address uses the first 8-bits of the 32-bit address to identify the network while the remaining 24 bits identify the host on the network. A Class B address uses the first 16 bits of the 32-bit address to identify the network address, while the next 16 bits identify the host on the network. The third address that can be subdivided is the Class C address. The first 24 bits of a Class C address identify the network while the remaining 8 bits of the 32-bit address identify the host on the network.

In comparison to Classes A, B and C that are subdivided into network and host portions Class D and E addresses are not subdivided. A Class D address represents a multicast address while a Class E address represents an experimental address. Figure 6.5 illustrates the five types of IP addresses.

Preamble (8 bytes)	Destination Address (6 bytes)	Source Address (6 bytes)	Type/ Length (2 bytes)	Data (46–1500 bytes)	CRC (4 bytes)

Figure 6.4 The Ethernet frame format

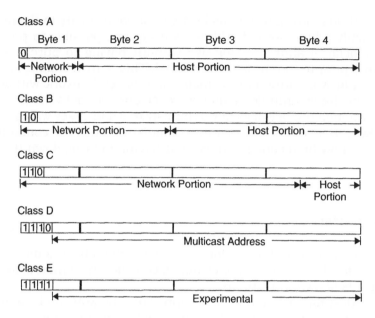

Figure 6.5 IPv4 address formats

The address resolution process

When an IP datagram arrives at a router connected to a LAN where the destination IP address resides, the router needs to place the datagram into one or more frames for delivery. However, because frames use the MAC address for data delivery, the router must determine the applicable MAC address associated with the destination IP address. This process is referred to as address resolution since the 32-bit IP address must be resolved into an applicable 48-bit MAC address.

The address resolution process begins by the router first examining its cache memory to determine if the address was previously resolved. If so, the router will extract the MAC address associated with the IP address to form a frame for delivery of the IP datagram via the LAN. If the IP address was not previously resolved, the router must determine the applicable MAC address. To do so the router will broadcast an address resolution protocol (ARP) frame onto the LAN. This frame will contain the IP address for which the router requires the corresponding MAC address. If the station with the destination IP address is powered on, it will recognize the ARP request and return its MAC address to the router. The router will place the MAC address in memory in association with the IP address as well as use the newly resolved address to create LAN frames for the delivery of the datagram.

Time considerations

The time required to resolve an address not only delays the delivery of data to the destination but, in addition, adversely affects the transmission of other data on the LAN. The reason for this is due to the fact that when the ARP broadcast and reply flows over the LAN this action precludes the ability of other stations to transmit information.

Using static ARP

One method you can consider to enhance the flow of network traffic is to use static ARP cache entries for enabling the router to translate 32-bit IP addresses into 48-bit hardware addresses. In doing so you could add static entries for a core set of popular servers and workstations so they are not periodically purged from router memory, resulting in ARP broadcasts being used to relearn the MAC addresses. The format of the 'ARP' command you would use to enter a permanent entry into the ARP cache is shown below:

```
arp<internet-address><hardware-address>type
```

Here the parameter `<internet-address>` represents the IP address configured on the workstation specified by the `<hardware-address>` parameter. The IP address is encoded in dotted decimal notation while the hardware address is encoded in triple quad-hex notation, with the latter representing a 48-bit value. The `<type>` argument defines the method of encapsulation. The value for 'type' is normally the keyword 'arpa' for an Ethernet network and is always 'snap' for FDDI and Token-Ring interfaces.

Configuration example

To illustrate an example of the configuration of a static ARP entry let us assume your organization's FTP server has the IP address 205.131.175.5. Let us further assume that the server is located on an Ethernet LAN and has the hardware address 0820.0910.A123. You would enter the following ARP command to place a static ARP entry in your router's cache.

```
arp 205.131.175.5 0820.0910.A123 arpa
```

Setting an ARP timeout

Another method you can consider to reduce the number of ARP broadcasts is to use the 'arp timeout' interface command to extend the time entries remain in a router's cache. By default, ARP entries are cached for 14400 seconds or four hours. The format of the 'arp timeout' command is shown below:

```
arp timeout <seconds>
```

If you enter a value of zero seconds, the cache entries are never cleared. Similar to viewing other parameters that may be of concern, you can use the 'show interfaces' command to display the ARP timeout value.

Although the use of static ARP entries and permanent cache entries may not appear to have an observable effect upon network performance, the old adage 'every bit helps' holds true, even if it sounds like a pun. That is, if we can literally shave a millisecond here and a millisecond there, such milliseconds will add up. When we consider the cumulative effect of all techniques covered in this chapter instead of their effect on an individual basis, we may obtain a significant increase in transmission performance that will enhance our overall QoS capability.

6.3 TAILOR THE ACCESS LINE

For a third QoS enhancement technique we will turn our attention to the access line that connects an organization's LAN to a network. That network can be a corporate intranet or the public Internet.

6.3.1 Overview

One of the major sources of transmission delay is the literal 'last mile' connection of an organization into the Internet. While a large portion of the Internet backbone consists of fiber optic transmission facilities operating at gigabit data rates, the copper connection typically used by most organizations to connect to the Internet operates at the Kbps to low Mbps data rate.

Packet delay

To obtain an appreciation of the minimum potential delay packets can experience entering and exiting the Internet, let us consider some common access rates. Today most organizations have a leased line connection to the Internet that operates at a data rate of at least 56 Kbps. If the communications carrier providing the digital connection has a modern digital infrastructure, its minimum data rate for a digital circuit over metallic wire is 64 Kbps. Thereafter, it is common for a communications carrier to offer fractional T1 service at operating rates that are a multiple of 64 Kbps up to 768 Kbps as well as a T1 operating rate of 1.544 Mbps. Concerning the latter, the T1 transmission facility uses 8 Kbps for framing, resulting in its data transmission capacity being reduced to 1.536 Mbps.

To illustrate the minimum potential delay associated with the operating rate of Internet access lines we need to consider the packet length since the packet length divided by the operating rate of the access line determines the ingress and egress delay. For illustrative purposes let us assume packet lengths of 72 and 1526 bytes. While these two metrics represent the minimum and maximum length Ethernet frames permitted to flow on a LAN, they also represent common Layer 3 packet lengths.

Table 6.2 Ingress/egress access line delay in seconds

Operating rate bps	Packet length	
	72 bytes	1526 bytes
56000	0.01028571	0.21800000
64000	0.00900000	0.19075000
128000	0.00450000	0.09537500
192000	0.00300000	0.06358333
256000	0.00225000	0.04768750
384000	0.00150000	0.03179167
448000	0.00128571	0.02725000
512000	0.00112500	0.02384375
576000	0.00100000	0.02119444
640000	0.00090000	0.01907500
704000	0.00081818	0.01734091
768000	0.00075000	0.01589583
1536000	0.00037500	0.00794792

Table 6.2 denotes the ingress and egress access line delays in fractions of a second at 13 distinct access-line operating rates.

To illustrate how the computations occurred let us examine the calculations for the delay associated with a 72-byte packet flowing via an access line at 56 Kbps. The time delay is computed as follows:

```
72 bytes X 8 bits/byte = .0102857 seconds
---------------------
56000 bits/second
```

Application considerations

In examining the entries in Table 6.2 it is important to note that the application being performed governs whether or not the access delay times will adversely affect the activity being performed. For example, consider an organization that is implementing a voice over IP (VoIP) application to provide voice communications between a regional office and a branch office. Let us assume that the branch office is connected to the Internet at 64 Kbps while the regional office has a T1 connection. If digitized voice packets are transported with an average length of 72 bytes, then the latency or delay at the ingress from Table 6.2 is. 010 seconds or 10 ms. At the egress point the delay is. 000375 seconds or. 375 ms. Thus, the total ingress and egress delay is 10.375 ms. While this delay may not appear significant, it is important to note that it represents approximately 7 percent of the total one way allowable delay of 150 ms prior to a conversation taking on the characteristics of a citizen band (CB) radio, where each party needs to use the term 'over' to signal to the other party that they have finished talking and the other party can now commence talking. As the latency or delay increases, one party to a conversation finds it difficult to determine when the other party finished speaking. Due to this, the other party may begin speaking only to hear the delayed arrival of speech, resulting in a bit of confusion and a back-off from speaking. Thus, any method that can reduce end-to-end latency to include increasing the operating rate of an access line deserves consideration.

6.3.2 Reducing latency

An organization running VoIP or a multimedia application between two locations needs to consider the access line operating rates at each ingress and egress location. By examining the overall network delay that can be obtained through the use of the PING application and noting the latency from egress and ingress lines, you can then determine the potential effect resulting from upgrading one or more access lines. If the end-to-end delay to include routing via an intranet or the Internet approaches or exceeds 150 ms, it is quite possible that an increase in the operating rate of one or more access lines may result in a marginal application becoming viable.

6.4 ENABLING RTP HEADER COMPRESSION

It is a well-known fact that data compression reduces the quantity of data that needs to be transmitted. As we reduce the amount of data that needs to be transmitted we not only reduce transmission time but in addition, reduce overall network traffic. In this section we will turn our attention to a specialized form of compression that is applied to a troika of headers.

The Real Time Protocol (RTP) represents an Internet Protocol that conveys timing information for the transport of real-time audio and video data.

6.4.1 Overview

Although the length of the RTP header is only 12 bytes, when you consider the fact that the transmission of digitized voice results in IP, UDP and RTP headers prefixed to a relatively small payload the header overhead becomes significant.

Advantages of header compression

The top portion of Figure 6.6 illustrates the formation of an IP datagram containing a payload that contains digitized voice. Note that the three headers cumulatively result in a prefix of 40 bytes to a payload that commonly varies from 20 to 160 bytes. Thus, the headers can represent between 66 percent overhead on a worst case basis to 20 percent overhead on a best case basis of the total length of the IP datagram.

If you configure RTP header compression, its effect will be to reduce the IP/UDP/RTP header from 40 bytes to approximately 2 to 4 bytes as illustrated in the lower portion of Figure 6.6. Not only does this action reduce the header overhead but, in addition, reduces the total amount of traffic that needs to flow over a communicative link. If your organization is using a relatively slow

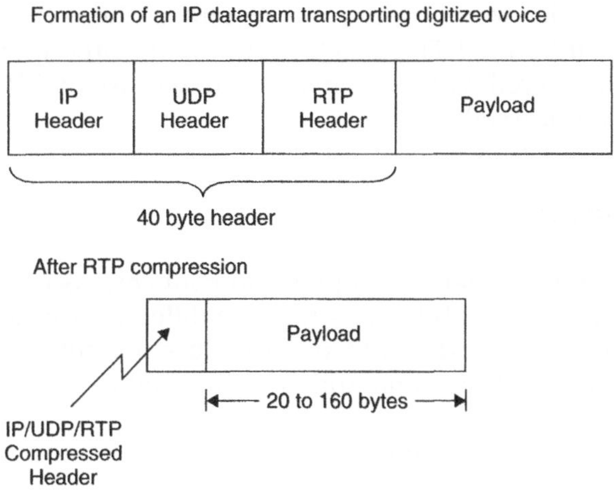

Formation of an IP datagram transporting digitized voice

IP Header	UDP Header	RTP Header	Payload

40 byte header

After RTP compression

	Payload

|←——— 20 to 160 bytes ———→|

IP/UDP/RTP
Compressed
Header

Figure 6.6   The potential effect of RTP header compression

digital transmission facility, the employment RTP header compression can have a significant effect upon reducing communications induced delay. This in turn enhances your organization's ability to obtain a desired QoS level.

Operation

The RTP header compression scheme employs the compressed Real-time Transport Protocol (RTP) to shrink the IP/UDP/RTP header from 40 bytes to 2 to 4 bytes most of the time. When no UDP checksums are employed, CRTP will reduce the tri-header to 2 bytes, while the use of UDP checksums results in CRTP shrinking the tri-header to 4 bytes. The ability of CRTP to literally shrink the tri-header results from the fact that there are only minor changes in several fields of the three headers and their difference from packet to packet is often constant, permitting compression to become very effective. This permits each end of a link to maintain the uncompressed header and the first-order differences in the session state shared between the router compressing the header and the router at the other end of the link that performs decompression. Then, all that needs to be communicated in addition to the payload is an indication that the second-order difference in the tri-headers was zero. This enables the router performing header decompression to reconstruct the original header by adding the first order differences to the uncompressed header as each packet is received.

6.4.2 Configuring RTP header compression

To enable RTP header compression you would use the 'ip rtp header-compression' command. The format of this command, which is applicable for a router's serial interface, is shown below:

```
ip rtp header-compression [<passive>]
```

The inclusion of the keyword <passive> results in the software compressing RTP packets only if inbound RTP packets on the same interface are compressed. If the command is entered without the keyword 'passive', all RTP traffic on the interface will be compressed.

Connection limitations

By default, up to 16 RTP header compression connections on an interface can be supported. To obtain the ability to support a different number of RTP header compression connections you would use the 'ip rtp compression connection' command. The format of this command is shown below:

```
ip rtp compression-connections <number>
```

This occurs where the variable <number> is used to specify the total number of RTP header compression connections that can be supported on a particular interface.

Configuration example

For our configuration example let us assume we want to enable RTP header compression on serial port Ø. Let us further assume we will use encapsulated PPP and want to support a maximum of 32 RTP header compression connections on the interface. To accomplish the previously mentioned items we would enter the following commands:

```
interface serial Ø
ip rtp header-compression
encapsulation ppp
ip rtp compression-connections 32
```

6.5 ENABLING OTHER COMPRESSION METHODS

Although RTP header compression can be used to significantly enhance transmission on relatively low speed serial connections, it is not the only data compression method supported on Cisco routers. Two additional data compression methods you can consider enabling are tcp header compression on ip and frame relay connections. Thus, in this section we will turn our attention to each compression method.

6.5.1 TCP header compression

Cisco routers provide support for compressing the headers of TCP/IP packets as a mechanism to reduce the size of such packets. To do so you would use the 'ip tcp header-compression' command on your router's serial interface on frame relay, HDLC or for Point-to-Point Protocol (PPP) encapsulation. The format of this command is shown below:

```
ip tcp header compression [<passive>]
```

Here the keyword <passive> results in the router compressing outbound TCP packets only if inbound TCP packets on the same interface are compressed. If this optional keyword is not specified, all TCP packets will be compressed.

Operation

The use of TCP header compression requires both ends of a serial connection to have this feature enabled. The actual compression method reduces the header which results in a lower overhead when short payloads are transmitted than when long payloads are transferred. This means that TCP header compression will be more effective for such applications as Telnet and conveying call-control signals than for file transfers. Because the enabling of TCP header compression disables a router's fast switching capability, this means that it is possible for a fast interface to overload your router. Due to this you should examine packet drops after you enable TCP header compression to ensure it does not result in an undesirable router overload.

Configuration example

To illustrate the use of the 'ip tcp header compression' command let us again assume we wish to enable such compression on the router's serial Ø port. We should also note that, similar to RTP header compression, ip tcp compression provides a command that can be used to specify the total number of header compression connections that can occur on an interface. That command is the 'ip tcp compression-connections' command that supports the suffix of a number that represents the maximum number of header compression connections that can exist on an interface. Based upon the preceding, the following commands enable tcp header compression on the serial Ø port with a maximum of ten header compression connections.

```
interface serial Ø
ip tcp header-compression
ip tcp compression-connections 10
```

6.5.2 Frame relay TCP header compression

As previously noted, we can enhance the performance of a frame relay circuit by transmitting TCP/IP headers in compressed form. To do so we must first configure frame relay on an applicable interface. Next, we would use the 'frame-relay ip tcp header-compression' command whose format is shown below:

```
frame-relay ip tcp header-compression [<passive>]
```

Similar to the other compression commands covered in this section, the key-word <passive> causes the outbound TCP/IP packet header to be compressed only if an incoming packet has a compressed header.

Configuration example

To illustrate the use of TCP header compression in a frame relay environment let us again assume we are using the serial Ø interface. Then, we would use the following commands to enable Cisco encapsulation and TCP header compression:

```
interface serial Ø
encapsulation frame-relay
frame-relay ip tcp header-compression
```

6.6 ELIMINATE DIRECTED BROADCASTS

In this section we will examine a technique whose implementation can not only reduce network delays but in addition enhance the security level of our organization's network. As we will shortly note, disabling directed broadcasts can deny hackers an easy weapon as well as reduce transmission on both our organization's network and WAN connection.

6.6.1 Overview

A broadcast represents a data packet that is directed to all hosts on a particular network. While we tend to consider the effect of broadcasts being local in scope, a special type of broadcasts can adversely affect both our organization's LAN and WAN. That type of broadcast is known as a directed broadcast.

IP broadcasts

Earlier in this chapter we examined the manner by which Class A, B, and C addresses are subdivided into network and host portions. We noted that the first 8 bits of the 32-bit Class A address denotes the network while the remaining 24 bits denotes the host on the network. Similarly, we noted that a Class B network address uses the first 16 bits of the 32-bit address to identify the network while the remaining 16 bits identify the host on the Class B network. Finally, we noted that the subdivision of the 32-bit Class C address

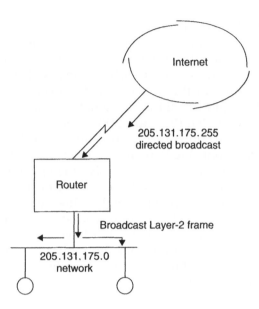

Figure 6.7 A directed broadcast represents a Class A, B, or C network address with its host address portion set to all 1's. A router connected to the network converts the IP address into a Layer-2 broadcast

results in the first 24 bits representing the network address while the last 8 bits represent the host on the network.

If a device sets the host portion of an IP address to all ones, this represents a broadcast address that applies to the network portion of the address. This type of broadcast is referred to as a directed broadcast and its occurrence can have an adverse effect upon your organization's LAN as well as its WAN connection to the Internet. To denote why directed broadcasts can literally clog your organization's LAN and WAN, consider Figure 6.7. In this example, let us assume your organization's IP network address is the Class C address 205.131.175.0. If we assume a packet with the IP address of 205.131.175.255 flows towards your network, your router will recognize the packet as a directed broadcast. This means that the router will create a MAC Layer 2 broadcast frame to transport the packet onto the network behind the router. This also means that each station on the LAN will read the frame.

Hacker abuse

While there are some situations where the use of a directed broadcast can represent a valid application, beginning a few years ago hackers began to abuse its use. For example, directing a ping to an IP broadcasts address results in each powered-on workstation on the network responding with an echo-reply message. Thus, we assume the network has 200 powered-on and thus active workstations, each ping would result in 200 responses. Because source IP addresses are not checked by routers, a common unsophisticated denial of service attack method is for a hacker to configure their workstation with a phony IP address and continuously ping the directed broadcast address of the network they wish to attack. If the hacker configures the source IP address of their workstation as the address on another network they wish to attack, the directed broadcast results in the first targeted network attacking the second! Even though the default ping is 32 bytes in length, a continuous ping generated ten times per second for which there are 200 stations replying could generated 32 bytes × 8 bits/byte × 10 times/sec × 200 stations or 512000 bits/second. If your organization was connected to the Internet via a T1 line operating at 1.544 Mbps just one workstation continuously pinging your directed broadcast address would result in the use of one-third of your outbound bandwidth. Now suppose the hacker is a student at a high school or college who goes into the computer laboratory and sets a dozen workstations to continuously ping your network's broadcast address. Instead of having your T1 circuit occupied only a third of the time with extraneous echo-replies it may now be filled with echo-replies, precluding or making it extremely difficult for legitimate users to receive a response to their queries.

6.6.2 Disabling directed broadcasts

Recognizing the threat of directed broadcasts Cisco changed the default of directed broadcasts enabled to being disabled in a recent release of its Internetwork Operating System (IOS). However, if you are using a version of IOS prior to 12.0 you should insure directed broadcasting is disabled. To do so you would enter the following IOS command:

```
no ip directed-broadcast
```

Based upon the preceding, the use of the 'no ip directed-broadcast' command can both enhance network performance as well as provide an additional level of security for your network. Due to this, directed broadcasts are now disabled by default in recent release of Cisco's IOS. However, if you are using an older version of the operating system it is a good idea to disable this broadcast capability.

6.7 ENABLE SELECTIVE ACKNOWLEDGEMENTS

In the previous section in this chapter we examined a QoS enhancement technique that results from the disabling of a router function. In this section we will take an opposite approach, examining how the enabling of selective acknowledgements can provide an enhanced data transmission capability that will in turn facilitate our QoS quest.

6.7.1 Overview

In a TCP environment it is possible to obtain a degraded level of performance when multiple packets are lost from one window of data. To understand how this can occur we need to review how TCP denotes the duration of a session.

TCP operations

As TCP is initiated it measures the round-trip time (RTT) between source and destination. The software module assumes that the TCP session will last for a number of RTTs so that the initial 3-way handshake does not adversely affect the application. During the TCP session numerous RTT intervals are computed which enable the software to determine the characteristics of the connection. When transmitting over a terrestrial circuit with a large number of router hops or via a satellite system, the delay-bandwidth product of a transmission path will be used to determine the amount of data TCP should send along the transmission path at any point in time to utilize the available transmission capacity. The delay used comes from the RTT, while the bandwidth represents the capacity of the bottleneck in the path between source and destination. As the delay increases, a TCP flow needs to place a larger amount of data on the transmission path. As the sender increases its window to operate on the path more efficiently, the probability of multiple packet drops per RTT window increases.

Using selective acknowledgements

To minimize the potential effect of the previously described situation you can enable TCP selective acknowledgements through the use of the 'ip tcp selective-ack' command. The format of this command is:

```
ip tcp selective-ack
```

Note that this command has no arguments nor keywords. Through the use of selective acknowledgements the receiving TCP device will return selective acknowledgements to the sender, in effect informing the sender about data that was received. This enables the sender to only have to retransmit missing data segments, which will result in an improvement of the overall efficiency of a TCP session.

Although enabling TCP selective acknowledgement can result in an enhanced flow of data when transmission occurs via satellite or on terrestrial circuits with a large number of router hops, it is important to note it is mutually exclusive with TCP header compression. Thus, you need to consider the type of transmission path to be used and its bit error rate prior to deciding to use one feature over another. To assist you in the feature selection process you can consider using the traceroute application which is called tracert in a Microsoft Windows environment. If you are using Windows 2000 you can also consider using pathping, a relatively new application that some people consider a combination of tracert and ping on steroids. The use of either traceroute (tracert) or pathping will enable you to determine the end-to-end round-trip delay between sender and receiver as well as the delays at each hop in the path. Based upon this information you can decide whether or not the enabling of selective acknowledgements is better than TCP header compression and then implement one or the other.

6.8 ENABLE LINK FRAGMENTATION AND INTERLEAVING

Earlier in this chapter we noted that the acceptable delay for transmitting voice over a data network is 150 ms. We also noted that we could increase the access line operating rate to reduce ingress and egress delays as well as perform a variety of other functions that reduce the quantity of data transmitted or make the movement of data more efficient. While the previously mentioned techniques in this chapter all deserve to be considered, their use does not preclude the ability of relatively long data transporting packets to randomly get between two packets transporting digitized voice. When this situation occurs, it is quite possible for the delay introduced by the need to output the large data packet to significantly exceed all or a large portion of the 150 ms end-to-end delay that the transport of voice can tolerate prior to a voice conversation deteriorating to a citizen band (CB) conversation.

6.8.1 The need for fragmentation

To illustrate the need for packet fragmentation let us assume a 1500-byte packet transporting data gets between two 64-byte packets transporting digitized voice. Let us further assume that the serial communications link connecting two organizational locations operates at 64 Kbps. Then the delay attributable to the 1500-byte data packet being serviced becomes:

```
1500 bytes x 8 bits/byte = 0.214s
------------------------
         56000 bps
```

Based upon the prior computation an organization would not be able to service data transporting packets when transmitting digitized voice if they wished to ensure packets were not adversely delayed. Obviously, this would not be a reasonable solution to this problem and this resulted in the development of link fragmentation and interleaving for use with several protocols and which is the topic to be discussed in this section.

6.8.2 Overview

In addition to real-time multi-media there are other types of traffic that can be adversely affected when relatively long packets become inserted between relatively short packets used by some applications. For example, consider Figure 6.8, which illustrates the flow of traffic onto a wide area network resulting from a Telnet and an FTP session occurring concurrently.

Note that periodically a relatively long FTP packet will gain access to the serial communications link, resulting in the long packet flowing between the packets generated by the Telnet session. This situation will occur even when queuing is employed since the FTP session must periodically be serviced. Thus, the only

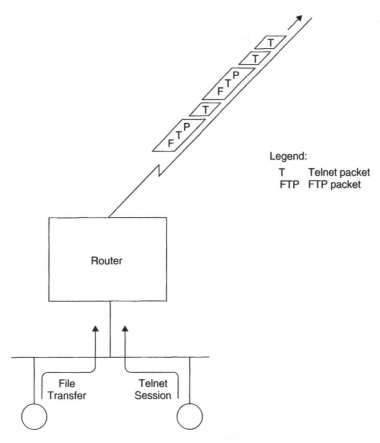

Figure 6.8 The need for fragmentation and interleaving occurs when relatively long packets are randomly inserted between time-dependent relatively short packets

method to minimize the delay resulting from relatively long packets inserted between time dependent packets is to fragment the long packets.

In a Cisco environment you can enhance link-layer efficiency through the employment of link fragmentation and interleaving (LFI). Currently link fragmentation and interleaving are supported for Multilink PPP (MLP), Frame Relay and Asynchronous Transfer Mode (ATM) Virtual Circuits.

6.8.3 Operation

Cisco's LFI feature can be used to reduce latency or delay on slow speed serial communications links. It reduces latency by breaking up lengthy datagrams and interleaving low-delay traffic packets with the smaller packets created from the fragmented datagram. Although Cisco's LFI feature follows the specifications of RFC 1717, it is important to note that to ensure the correct order of transmission and re-assembly of fragmented packets, each packet must be prefixed with an applicable header. Due to this you can expect your total link utilization to increase slightly, this is the price you can expect to pay for minimizing the delay that can result from long packets being interleaved between time-dependent packets.

Now that we have an appreciation of LFI let us turn our attention to the manner by which we would configure this feature for different multiple logical links.

Multilink PPP

Cisco's Link Fragmentation and Interleaving (LFI) feature supports the Multilink Point-to-Point Protocol (MLP), where MLP provides a mechanism for splitting, sequencing and recombining datagrams so that they can flow across multiple logical data links. The ability to interleave data is only supported on certain interfaces, with ISDN and dialer interfaces the most common types of interfaces supported. The actual configuration of MLP interleaving represents a five-step process. First, you need to enable MLP. To do so you would enter the following interface command:

```
ppp multilink
```

The second step in the MLP interleaving configuration process is to enable real-time packet interleaving. To accomplish this task you would enter the following interface command:

```
ppp multilink interleave
```

The third step in the MLP configuration process is to configure a maximum fragment delay. This step is optional and results in the selection of a fragment length by MLP that ensures that the specified delay is not exceeded. To accomplish setting a maximum fragment delay you would enter the following interface command:

```
ppp multilink fragment-delay <milliseconds>
```

This occurs where the parameter <milliseconds> denotes the maximum outbound delay.

Continuing our MLP interleaving configuration process, the fourth step is to reserve a queue for real-time packet flows to a specified UDP port. This step is also optional, however, if running VoIP, its use is highly recommended. This is because its use allows you to allocate real-time traffic to a higher priority than other traffic flows. You would use the 'ip rtp reserve' command to reserve a special queue for real-time packet flows to specified UDP ports. The format of this command is:

```
ip rtp reserve <lowest-UDP-port><range-of-
    ports>[<maximum-bandwidth>]
```

The specification of a maximum bandwidth is optional. If specified and the use of actual bandwidth exceeds the specified limit, the reserved queue will be degraded to provide a best-effort queue capability.

The fifth and final step in the MLP interleaving configuration process is only applicable for virtual interface templates. This step involves applying the virtual interface template to the multilink bundle. To do so you would enter the command 'multilink virtual-template' without any parameters or keywords.

Configuration example

To illustrate the configuration of LFI for MLP we need to have an example. Let us assume we want to define a virtual interface template that will configure MLP interleaving. In doing so we will also assume we want to configure a maximum real-time traffic delay of 20 ms. Last but not least, we wish to reserve a special queue for packets flowing to UDP ports 32768 through 32777. To accomplish the preceding as well as apply the virtual template to the MLP bundle we would enter the following commands:

```
interface virtual-template 1
ip unnumbered ethernet Ø
ppp multilink
ppp multilink interleave
ppp multilink fragment-delay 20
ip rtp interleave 32768 10
multilink virtual-template 1
```

LFI for frame relay and ATM

Similar to providing link fragmentation and interleaving (LFI) for MLP, Cisco also supports this capability for frame relay and ATM virtual circuits (VCs). In this section we will first focus our attention upon configuring LFI using the Multilink Point-to-Point Protocol (MLP) over frame relay. Once this is accomplished, we will turn our attention to LFI over ATM VCs.

Configuring LFI for frame relay

To configure LFI using Multilink Point-to-Point Protocol (MLP) over frame relay you need to first configure LFI using MLP in a virtual template interface (VTI). Once this is accomplished you then need to associate the VTI with a frame relay permanent virtual circuit.

The configuration of LFI using MLP in a virtual template interface represents a six-step process. Table 6.3 lists the applicable router commands as well as providing a brief description of the use of each command.

Once you configure LFI using MLP in a virtual template interface you need to associate the VTI with a frame relay PVC. The association of the VTI with a frame relay PVC requires you to enter four commands, beginning in the global

Table 6.3 Configuring LFI using MLP in a virtual template interface

Command	Description
`interface virtual-template`<number>	Create a virtual template and enter interface configuration mode.
`bandwidth`<kilobits>	Set bandwidth for an interface.
`service-policy output`<policy-name>	Attach the specified policy map to the output interface.
`ppp multilink`	Enable MLP on the interface.
`ppp multilink fragment-delay`<milliseconds>	Configure the maximum delay allowed for the transmission of a packet fragment.
`ppp multilink` interleave	Enable interleaving of RTP packets among fragments.

configuration mode. First, you need to configure an interface type. To do so you would enter the following command:

> `interface`<type number>

The second step in the association process is to enable frame relay traffic shaping on the interface. To do so you would enter the following command:

> `frame-relay traffic-shaping`

The third step is to associate a virtual template interface with a frame relay data link connection identifier (DLCI). To do so, you would enter the following command:

> `frame-relay interface-dlci`<dlci>[`ppp`<virtual-template-name>]

The fourth and final step in the association process is to associate a frame relay map class with a DLCI. To do so, you would enter the following command:

> `class`<name>

Configuring LFI using MLP over ATM

Similar to configuring LFI using MLP over frame relay, the configuration process for ATM represents two distinct tasks. First, you need to configure LFI using MLP on a virtual template interface. The steps to perform this operation are the same as that previously described for frame relay and were listed in Table 6.3. However, unlike frame relay that can have variable length frames, ATM uses fixed length cells that consist of a 5-byte header and 48-byte payload. Thus, the selection of a fragment size for MLP over ATM should enable the fragments to fit into an exact multiple of ATM cells. According to Cisco literature, the fragment size in bytes for MLP over ATM is calculated using the following formula:

```
Fragment size = 48 x number of cells - 10
```

While the subtraction of 10 may appear questionable, it should be noted that there are several types of frame relay connections with some requiring the use of one or more bytes in the payload section. Thus, the subtraction of 10 bytes is more than likely intended to compensate for the use of one or more bytes in the payload portion of each cell.

Once you configure LFI using MLP on a virtual template interface, the second task is to associate the VTI with an ATM PVC. To do so you need to perform a sequence of four steps.

First, you need to specify the ATM interface type. To do so you would enter one of the following commands, with the command entered based upon the ATM hardware used in your router:

interface atm slot/∅
or **interface atm** slot/port

The second step requires you to create an ATM PVC. To do so you would enter the following command:

pvc [<name>] <vpi/vci>

Here the vpi represents the virtual path identifiers while the vci represents the virtual channel identifier.

The third step in the association process is to select the available bit rate (ABR) QoS and configure the output peak cell rate (pcr) and output minimum guaranteed cell rate (mcr) for an ATM PVC. To do so, you would enter the following command:

abr <output-pcr><output-mcr>

The fourth and final step in the association process is to specify that PPP is established over the ATM PVC using the configuration from the specified virtual template. To do so, you would enter the following command:

protocol ppp virtual-template<number>

MONITORING YOUR NETWORK

In this concluding chapter we will focus our attention upon obtaining information that can be of considerable value in making decisions concerning the adjustment of the allocation of network resources. In doing so we will focus our attention upon the use of the 'show' command, using different keywords with that command to tailor a display to our specific requirements. Because the results of the use of various show commands may require an adjustment in the configuration of hardware, when applicable we will also examine the use of various configuration commands.

7.1 THE SHOW COMMAND

In actuality there is no one 'show' command. Instead, there are literally hundreds of show commands supported by a Cisco router. Such show commands are normally followed by a keyword which identifies the type of data to be displayed and may optionally be followed by one or more parameters.

Table 7.1 lists 23 examples of the use of the 'show' command to include a brief description of the display resulting from the use of each command.

In addition to the variations of the 'show' command included in Table 7.1 there are numerous other 'show' commands that support specific protocols and can be used with many keywords to display specific protocol related information. For example, you can use the 'show' command with such protocol keywords as 'appletalk', 'decnet', 'frame-relay' as well as router protocols to include the border gateway protocol (bgp), the exterior gateway protocol (egp) and open shortest path first (ospf) protocol. When you consider that each of the show commands used to display information about different protocols can support a number of keyword modifiers, the result is a universe of several hundred show commands. Because this book is focused upon Quality of Service, which is primarily applicable to an IP environment, the majority of our discussion concerning the use of the show command will actually involve variations of the 'show ip' command. However, because most organizations commonly support a mixture of protocols on their internal network, we will commonly need to examine the cumulative effect of all protocols on network and router performance.

Table 7.1 Examples of the 'show' command

Command	Description
show arp	Displays entries in the ARP table of the router.
show cls	Displays the current status of all Cisco link services (CLS) sessions on the router.
show context	Displays NVRAM information when the router crashes.
show debugging	Displays information about the type of debugging enabled for the router.
show diag	Displays hardware information on the line cards.
show environment	Displays temperature, voltage and blower information for certain routers and switches.
show hosts	Displays default domain names.
show ip accounting	Displays active accounting information.
show ip arp	Displays ARP cache where SLIP addresses appear as permanent ARP table entries.
show ip cache	Displays routing table cache used to fast switch IP traffic.
show ip interface	Displays the status of an interface.
show ip redirects	Displays the address of a default gateway (router) and the address of hosts for which a redirect was received.
show ip route	Displays entries in the routing table.
show ip route summary	Displays summary information about entries in the routing table.
show ip tcp header-compression	Displays statistics about TCP header compression.
show ip traffic	Displays statistics about IP traffic.
show logging	Displays the state of logging.
show memory	Displays statistics about memory utilization.
show processes	Displays information about active processes.
show processes memory	Displays information about memory used.
show protocols	Displays information about the global and interface-specific status of any Layer 3 configured protocol.
show tcp	Displays the status of TCP connections.
show tcp brief	Displays a concise description of TCP endpoints.

Thus, we will need to display additional information beyond the IP protocol and will do so through the use of other versions of the show command.

7.1.1 User versus privilege mode

Prior to actually working with applicable show commands it is probably a good idea to review the two router access modes of operation – user and privileged and their relationship to the command interpreter, the latter referred to as the EXEC.

The EXEC

The command interpreter (EXEC) is responsible for interpreting commands entered into the router. Known as the EXEC, the command interpreter checks each command and, assuming they are correctly entered, performs the required operation.

If a router administrator entered a password during the set-up process, you must log into the router and provide the correct password prior to using an EXEC command. In actuality two passwords can be required to be entered to use EXEC commands, as there are two EXEC command levels. Those two levels are user and privileged.

User mode

When you initially log into a Cisco router you are placed in the user mode of operation. If a password was assigned to the user mode during the router set-up process you would first enter the applicable password to enter this mode. Once in the user mode, the system prompt in the form of an angle bracket (>) will be displayed.

In the user mode of operation you can use the show command to display information about the router, provide a name to a logical connection, change the parameters of a terminal, and initiate other noncritical operations. However, you cannot configure the router. To obtain the ability to configure a router you must be in its privileged EXEC mode of operation.

Privilege mode

A second mode of operation supported by Cisco routers is referred to as the privileged mode. If the enable-password command was previously used to place a password block on accessing the router's privileged EXEC mode of operation, its entry becomes a two-step process. First, from the angle bracket (>) known as the user mode prompt you need to enter the command 'enable'. This action will result in the router prompting you for the applicable password. Once you enter the correct password the router prompt changes to a hache sign (#) which serves as an indication that you are in the privileged EXEC mode of operation.

Figure 7.1 illustrates the use of the enable command from the user EXEC mode followed by a response to the password prompt to obtain access to the router's privileged EXEC mode.

In this example the name assigned to the router is shown as 'Macon'. Note that after we entered the privileged EXEC mode the prompt changed to the hache (#) sign. At this point we entered the 'configure' command followed by the question mark (?) to display a list of configuration commands supported by the router we are using. Due to space constraints only a partial listing of the resulting list of configuration commands supported by the router were included in Figure 7.1. It should be noted that you could use the 'show' command to display most types of information in both the user and privileged modes of operation. There are certain exceptions to this duality, one being the ability to display passwords, which requires you to be in the privileged mode of operation.

Now that we have an appreciation of the difference between user and privileged modes, let us turn our attention to the primary focus of this chapter which is upon monitoring our network.

```
Macon>enable
Password:
Macon#configure
Configuring from terminal, memory, or network [terminal]?
Enter configuration commands, one per line. End with CNTL/Z.
Macon(config)#?
Configure commands:
  aaa                        Authentication, Authorization and Accounting.
  access-list                Add an access list entry
  alias                      Create command alias
  alps                       Configure Airline Protocol Support
  appletalk                  Appletalk global configuration commands
  arap                       Appletalk Remote Access Protocol
  arp                        Set a static ARP entry
  async-bootp                Modify system bootp parameters
  autonomous-system          Specify local AS number to which we belong
  banner                     Define a login banner
  boot                       Modify system boot parameters
  bridge                     Bridge Group.
  bstun                      BSTUN global configuration commands
  buffers                    Adjust system buffer pool parameters
  call-history-mib           Define call history mib parameters
  cdp                        Global CDP configuration subcommands
  chat-script                Define a modem chat script
  clock                      Configure time-of-day clock
  config-register            Define the configuration register
  controller                 Configure a specific controller
  decnet                     Global DECnet configuration subcommands
  default                    Set a command to its defaults
  default-value              Default character-bits values
  dialer                     Dialer watch commands
  dialer-list                Create a dialer list entry
  dlsw                       Data Link Switching global configuration commands
  dnsix-dmdp                 Provide DMDP service for DNSIX
  dnsix-nat                  Provide DNSIX service for audit trails
  downward-compatible-config Generate a configuration compatible with older
                             software
  dspu                       DownStream Physical Unit Command
  dss                        Configure dss parameters
  enable                     Modify enable password parameters
  end                        Exit from configure mode
  endnode                    SNA APPN endnode command
  exception                  Exception handling
  exit                       Exit from configure mode
  file                       Adjust file system parameters
  frame-relay                global frame relay configuration commands
  help                       Description of the interactive help system
  hostname                   Set system's network name
  interface                  Select an interface to configure
  ip                         Global IP configuration subcommands
  ipc                        Configure IPC system
```

Figure 7.1 Using the enable command to enter the privileged EXEC command mode and issue a router 'configuration' command

7.1.2 The show interface command

One of the first commands you can consider using to obtain an indication of overall network events is the 'show interface' command. You can use this command as-is, to display information about the status of all router interfaces or you can follow the command with a specific interface to tailor the display so that information about a single interface is shown. Concerning the display of specific interface information, the manner by which an I/O port is fabricated governs the manner by which you specify an interface.

Interface specification methods

If a port is built into a router, it is referenced directly by its number. For example, serial port Ø would be referenced in an interface command as follows:

```
interface serial Ø
or interface sØ
```

If a group of ports is fabricated on a common adapter card for insertion into a slot within a router, the reference to the port requires you to specify both the slot number and the port number. For example, on a Cisco 7200 or 7500 series router you would use the following format to specify a particular serial port:

```
Interface serial slot/port#
```

A variation of the above format worth noting occurs on Cisco 7200 series equipment. Cisco 7200 routers can have multiple ports fabricated on a port-adapter card. In addition, multiple port adapters can reside in a slot. In this hardware configuration you would need to use the following command format to reference a specific serial port:

```
Interface serial slot#/port adapter/port#
```

Using the show interface command

To illustrate the use of the show interface command let us apply it to an interface. Figure 7.2 illustrates the use of the show interface command applied to the serial 1 port on port adapter 0, which is installed in slot Ø in a router used by this author. Through the examination of the resulting display we can learn a considerable amount of information concerning the WAN connection.

General information

The first three lines displayed in Figure 7.2 can be considered as providing general information about the state of the interface. The first line in the resulting display informs us that both the hardware and the line protocol are operational. The second line in the display that begins with the term 'Hardware' informs us of the hardware used. This is followed by a description of the transmission facility that was originally entered during the configuration

```
Macon#show interface serial0/0/1
Serial0/0/1 is up, line protocol is up
  Hardware is cyBus Serial
  Description: MCI MGBC673F00020002
  Internet address is 4.0.156.6/30
  MTU 1500 bytes, BW 1544 Kbit, DLY 20000 usec,
      reliability 255/255, txload 192/255, rxload 18/255
  Encapsulation HDLC, crc 16, loopback not set
  Keepalive set (10 sec)
  Last input 00:00:09, output 00:00:00, output hang never
  Last clearing of show interface counters 23:23:04
  Input queue: 0/75/0 (size/max/drops); Total output drops: 55700
  Queuing strategy: weighted fair
  Output queue: 0/1000/64/55700 (size/max total/threshold/drops)
      Conversations 0/70/256 (active/max active/max total)
      Reserved Conversations 0/0 (allocated/max allocated)
  5 minute input rate 110000 bits/sec, 145 packets/sec
  5 minute output rate 1163000 bits/sec, 169 packets/sec
     7236335 packets input, 735289876 bytes, 0 no buffer
     Received 0 broadcasts, 0 runts, 1 giants, 0 throttles
     501 input errors, 499 CRC, 0 frame, 0 over-run, 0 ignored, 2 abort
     8537108 packets output, 2885916747 bytes, 0 under-runs
     0 output errors, 0 collisions, 0 interface resets
     0 output buffer failures, 4625730 output buffers swapped out
     2 carrier transitions
     RTS up, CTS up, DTR up, DCD up, DSR up
```

Figure 7.2 Using the show interface command to display information concerning the status of a router's serial port

process. In this example we can note that the serial port is connected to a particular MCI circuit.

IP address

Continuing our examination of the display of the serial interface shown in Figure 7.2, the fourth line denotes the IP address assigned to the serial port. Note that the forward slash (/) followed by the number 30 directly indicates the extended network portion of the IP address. Because the first digit of the IP address in dotted decimal notation has the value of 4, then the bit composition of the first byte is 00000100. Since the first bit position in the IP address is a zero, this indicates that the address is a Class A address.

If we remember the structure of a Class A network, the first byte represents the network address while the next three bytes represent the host position on the network. Thus, the /30 prefix indicates that the network portion of the IP address is extended from 8 to 30 bits internally.

MTU, bandwidth and delay

The next line, which begins with the term 'MTU' indicates the maximum transmission unit, which is set to 1500 bytes. The following term of 'BW' indicates the bandwidth or operating rate of the circuit connected to the serial interface, which is 1544 Kbit. That operating rate represents the speed of a T1 line. The third term on the line, DLY, indicates the delay, which is set to 20000 usec.

Performance indicators

The next line which is indented displays three fractions in terms of 255ths for reliability, txload and rxload, the latter two representing transmit load and receive load. These three fractions, as we will shortly note, represent performance indicators covering transmission reliability, transmission occupancy and receive occupancy.

The reliability metric indicates, as you might expect, the reliability of the interface. In this example a value of 255/255 means that during a 5-minute period it was 100 percent reliable. The txload metric, which is indicated in Figure 7.2 as having a value of 192/255 indicates the exponential average occupancy of the transmit side of the interface for a 5-minute period. A value of 255/255 would indicate full occupancy, while a value of 0/255 would indicate no transmit traffic. Thus, the value of 192/255 means that on average 192/255 or approximately 75 percent of the transmit capacity of the T1 line was being used in the past 5-minute period. Because the T1 line operates at 1.544 Mbps but supports a data rate of 1.536 Mbps due to 8 Kbps of framing, we can say the outbound or transmit traffic was equivalent to 1.536 Mbps x. 75 or 1.152 Mbps during the past 5 minutes.

For the rxload metric you will note a value of 18/255 which represents approximately 7 percent. This means that on average over the past 5 minutes 7 percent of the 1.536 Mbps capacity of the line or 107 Kbps flowed in the inbound or receive/direction. If you are puzzled as to the discrepancy between the occupancy of the transmit and receive sides of the T1 line connected to the serial interface, you need to consider the manner by which Web traffic flows. That is, a relatively short packet or datagram in the form of a URL request will return a relatively long series of packets or datagrams that transport the response in the form of one or more Web pages. Because the router examined by this author was connected to a LAN on which a popular Web server resides, the rxload primarily represents packets containing URLs requesting Web pages. In comparison, the txload metric primarily represents Web pages flowing out of the router's serial port into the Internet in response to URL requests. Thus, the transmission of delay sensitive traffic over the serial port into the Internet should be prioritized over Web traffic.

Encapsulation and loopback

The line in Figure 7.2 that begins with Encapsulation denotes that the protocol transported on the link is HDLC using CRC-16 error checking. On that same line we note that the port is not in its loopback mode.

Although not directly related to QoS, when the port is in the loopback mode you obviously cannot have any quality of transmission. Thus, a quick check of the status of many interface indicators to include the loopback status through the use of the show interface command may show you an easy method to correct what might otherwise appear to be a major problem.

Keepalive

The keepalive signal is a common method used to inform a distant party that in the absence of actual data the connection is to remain 'up'. To do this, each party to the conversation generates a periodic keepalive signal. Thus, a transmitted keepalive becomes a received keepalive.

The following line notes that every 10 seconds a keepalive signal will be transmitted to inform the router on the other end of the connection that in the absence of data the line connection is still up or active.

Last bit received and counter status

The line beginning 'Last input' denotes when the last bit was received and output. The time at which the last input and output bits occurred provides a mechanism to indicate if and when actual line activity occurred.

The following line informs us of the last clearing of the counters that maintain statistics.

Queue indicators

The following line that begins with the term 'Input queue' indicates the number of packets in the input queue. Each number is followed by a slash which indicates in order the number of packets currently in the queue (which at the time of the display was Ø), the maximum queue capacity (which in this example is 75 packets) and the number of packets that were dropped (which during the 5-minute period was zero). To the right of the input queue troika of information is a summary of total output drops, which in Figure 7.2 totaled 55700 packets. Note that from the prior line the last clearing of the show interface counters occurred almost 24 hours prior to this display. Thus, the total number of packets that were dropped represent drops over an approximate 24-hour period.

If you note from periodic monitoring on the use of third part software, such as the Multi-Router Traffic Grabber (MRTG), that the size of the queue is above zero and packet drops become significant, this can be used as an indication that the data rate of the access line is not sufficient for your applications. If the access line is provided by an Internet Service Provider, another useful tool is the line utilization graph that many service providers maintain and upon request will provide for their customers. Typically, the service provider can provide line utilization graphs that will indicate the level of utilization based on 5-, 10-, or 15-minute samples over a day, week, or month. While you can obtain similar statistics using MRTG or a similar third party product, many times a call to your ISP can provide line utilization information that might otherwise require a considerable period of time to develop.

Returning again to Figure 7.2, we note the display of the queuing strategy configured for the serial port. For this example weighted fair queuing is in place. Next, the output queue metrics that indicate the size of the queue, its maximum capacity, threshold and drops are denoted. Indented below the output queue information is a summary of total and reserved conversations. This is followed by a 5-minute summary of the input and output data rate in terms of bps and packets per second. Note that the 5-minute input rate is shown as 110000 bps, which, as previously explained, results from the fact that most input traffic consists of URLs requesting Web pages.

Limitations to note

When examining the statistics generated by the use of the show interface command you need to consider the application or applications you are operating or intend to operate in conjunction with the data flow you are experiencing. For example, if you were considering implementing a real-time application, the fact that the output rate is 75 percent of the line capacity can be obtained from the show interface command. However, you cannot determine the average packet length from the use of this command, which means that if you need to determine if fragmentation could be helpful, you would have to use a protocol analyzer to gather statistics on packet length. Thus, many times you will need to supplement your use of router commands with other monitoring tools to determine if the use of one or more router features represents a viable solution or even an improvement to an existing situation.

Runts and giants

The last portion of the display shown in Figure 7.2 provides a summary of statistics that are for the most part self-explanatory. However, if you carefully examine the entries you would more than likely be puzzled by the inclusion of statistics for runts on a serial port. As a refresher, a runt represents a packet discarded because it is smaller than the medium's minimum packet length. While we tend to associate runts with Ethernet frames that are less than 72 bytes in length (64 bytes when the preamble is not counted), it is also possible to have an HDLC runt. Similarly, the giants field which represents packets that are discarded because they exceed the medium's maximum packet length are also commonly associated with Ethernet. However, they are also applicable to other protocols. In fact, you will note in Figure 7.2 that one giant packet was encountered.

Over-run and ignored

Another two fields in the lower portion of Figure 7.2 that deserve mention are over-run and ignored. The over-run field indicates the number of times the serial receiver hardware was unable to hand received data to a hardware buffer as a result of the input rate exceeding the receiver's ability to service the data. By comparison, the ignored field indicates the number of received packets ignored by the interface hardware as a result of internal buffers running low. Typically broadcast storms and bursts of noise can result in an entry for the ignored field.

Now that we have an appreciation of the use of the show interface command, let us turn our attention to the use of a series of show commands that can be used to display information concerning TCP/IP. Thus, in the second section in this chapter we will turn our attention to the use of IP-related show commands.

7.2 IP-RELATED SHOW COMMANDS

In this section we will focus our attention upon the use of various versions of the IP show command as a mechanism to denote the activity occurring on an organization's IP network. Because the actual number of IP show command options can vary based upon the router you are using and the version of the operating system in use, it is a good idea to determine the IP show commands supported. Thus, let us literally begin at the beginning and turn our attention to displaying the IP show commands supported by our router.

7.2.1 Displaying the IP show commands

Similar to the display of other help or assistance information in a Cisco environment we need to use the question mark (?) to display a list of show ip commands. Thus, in the user or privileged mode we would enter the following command:

```
show ip ?
```

An example of the use of the show ip ? command used on the author's router is shown in Figure 7.3.

```
Router#show ip
accounting<checkpoint>    Accounting statistics
arp                       IP ARP table
bgp<address>              Border Gateway Protocol
cache                     Fast Switching cache
egp                       EGP peers
interface<name>           Interface settings
protocols                 Routing protocols
route<network>            Routing table
tcp<keyword>              TCP information, type 'showiptcp?' for list
traffic                   Traffic statistics
```

Figure 7.3 Displaying a list of available commands for monitoring the status of an IP network

When examining the entries shown in Figure 7.3 note that you can drill down further via the show ip command to obtain specific information for one or more of the higher level entries. For example, entering the 'show ip tcp ?' command would result in the display of a list of available TCP commands.

Now that we have an appreciation of the use of the show ip ? command to determine the EXEC commands supported by our router for monitoring activity on an IP network, let us turn our attention to the display of specific information.

7.2.2 Displaying the ARP cache

To display the contents of the ARP cache at any point in time you would use the following EXEC command:

```
show ip arp
```

As a result of the use of the above command, the contents of the ARP cache are displayed. Figure 7.4 illustrates an example of the potential contents of the ARP cache.

When examining the entries in Figure 7.4 note that the column labeled 'Protocol' denotes the protocol for the network address listed in the column labeled 'Address'.

The network address corresponds to the hardware address listed in the column with that heading. The column labeled 'Age' denotes the length of time in minutes of the cache entry, with a dash (-) used to indicate a recent entry placed in the ARP cache. The column labeled 'Type' indicates the type of ARP. Permissible entries include ARPA for Ethernet-type ARP, SNAP for RFC 1042 ARP, and Probe for the HP Probe Protocol. The sixth and last column which is labeled 'Interface' indicates the router interface where the address resolution process occurred.

```
Protocol   Address          Age (min)   Hardware Addr    Type    Interface
Internet   198.108.1.140    127         aa00.0400.6408   ARPA    Ethernet0
Internet   198.108.1.111    116         0800.2007.8866   ARPA    Ethernet0
Internet   198.108.1.115    23          0000.0c01.0509   ARPA    Ethernet0
Internet   192.31.7.24      5           0800.0900.46fa   ARPA    Ethernet1
Internet   192.31.7.26      21          aa00.0900.6508   ARPA    Ethernet1
Internet   192.31.7.27      -           aa00.0900.0134   ARPA    Ethernet1
Internet   192.31.7.28      57          0000.0c00.2c83   ARPA    Ethernet1
Internet   192.31.7.17      37          2424.c01f.0711   ARPA    Ethernet1
Internet   192.31.7.18      14          0000.0c00.6fbf   ARPA    Ethernet1
Internet   192.31.7.21      104         2424.c01f.0715   ARPA    Ethernet1
Internet   198.108.1.33     25          0800.2008.c52e   ARPA    Ethernet0
Internet   198.108.1.55     14          0800.200a.bbfe   ARPA    Ethernet0
Internet   198.108.1.6      69          aa00.0400.6508   ARPA    Ethernet0
Internet   198.108.7.1      9           0000.0c00.750f   ARPA    Ethernet3
Internet   198.108.1.1      -           aa00.0400.0134   ARPA    Ethernet0
```

Figure 7.4 Displaying the contents of the ARP cache

By carefully examining the entries in the ARP cache against devices on your network you can decide whether one or more entries should be changed. As previously noted earlier in this book, under certain conditions you may wish to consider placing permanent entries in the ARP cache. Doing so both reduces ARP broadcasts as well as the small delay time associated with the address resolution process.

7.2.3 Displaying IP accounting

Continuing our tour of the use of the general show ip command, let us turn our attention to the show ip accounting command whose use allows us to display the active accounting database. This action will result in the display of four columns of data as illustrated in Figure 7.5.

Entries in the first two columns, which are labeled 'Source' and 'Destination', indicate the addresses of the originator and recipient, respectively. Entries in the next two columns, which are labeled 'Packets' and 'Bytes', indicate the total number of packets and bytes transmitted between each address pair.

When examining the sample output in Figure 7.5 resulting from the use of the show ip accounting command, note you can easily determine the distribution of activity by address pairs. In addition, since both packets and bytes transmitted are indicated, you can easily obtain the average packet length by dividing the former by the latter. If you have knowledge of the activities performed by

```
Source          Destination    Packets      Bytes
198.108.19.40   192.68.68.20         8        306
198.108.13.55   192.68.68.20        68       2849
198.108.2.50    192.12.33.51        18       1111
198.108.2.50    192.93.2.1           5        319
198.108.2.50    192.93.1.2         463      30991
198.108.19.40   192.93.2.1           4        262
198.108.19.40   192.93.1.2          28       2552
198.108.20.2    128.18.6.100        39       2184
198.108.13.55   192.93.1.2          35       3020
198.108.19.40   192.12.33.51      1986      95091
198.108.2.50    192.68.68.20       233      14908
198.108.13.28   192.68.68.53       390      24818
198.108.13.55   192.12.33.51    214669    9806659
198.108.13.111  128.18.6.23      28839    1126608
198.108.13.44   192.12.33.51     35412    1523980
192.31.8.21     192.93.1.2          11        824
198.108.13.28   192.12.33.2         21       1862
198.108.2.166   192.31.8.130       898     141054
198.108.3.11    192.68.68.53         4        246
192.31.8.21     192.12.33.51     15696     695635
192.31.8.24     192.68.68.20        21        916
198.108.13.111  128.18.10.1         16       1138
```

Figure 7.5 An example of the traffic statistics generated by the use of the show ip accounting command

the source addresses that reside on your network, the use of the show ip accounting command coupled with some minor mathematics can inform you if fragmentation and interleaving would provide a potential benefit to facilitate the flow of real-time data.

7.2.4 Displaying host statistics

Again continuing our journey covering the monitoring of an IP network, we can use the show hosts command to display information concerning the default domain name, the style of name look-up service, a list of name server hosts, and the cached list of host names and addresses. While this information can be useful in performing an analysis of the name look-up process, it normally will not affect the ability to obtain a QoS capability. Due to this it is left to the reader to probe further into the use of the show hosts command.

7.2.5 Displaying the route cache

Through the use of the show ip cache command you can display the contents of the routing table cache that is used to rapidly switch IP traffic. Entering the show ip cache command at the EXEC prompt will result in the display of the IP routing cache similar to the cache displayed in Figure 7.6.

```
IP routing cache version 435, entries 19/20, memory 880
Hash     Destination     Interface    MAC Header
*6D/0    205.18.1.254    Serial0      0DE00800
*81/0    198.108.1.111   Ethernet0    0000D002C83AC00040002340800
*8D/0    198.108.13.111  Ethernet0    AC0004000134AC00040002340800
99/0     205.18.10.1     Serial0      0DE00800
*9B/0    205.18.10.3     Serial0      0DE00800
*B0/0    205.18.5.39     Serial0      0DE00800
*B6/0    205.18.3.39     Serial0      0DE00800
*C0/0    198.108.12.35   Ethernet0    AC0004000134AC00040002340800
*C4/0    198.108.2.41    Ethernet0    0000D002C83AC00040002340800
*C9/0    192.31.7.17     Ethernet0    2424C01F0711AC00040002340800
*CD/0    192.31.7.21     Ethernet0    2424C01F0715AC00040002340800
*D5/0    198.108.13.55   Ethernet0    AC0004006508AC00040002340800
*DC/0    224.93.1.2      Serial0      0DE00800
*DE/0    192.12.33.51    Serial0      0DE00800
*DF/0    198.108.2.50    Ethernet0    AC0004000134AC00040002340800
*E7/0    198.108.3.11    Ethernet0    0000D002C83AC00040002340800
*EF/0    192.12.33.2     Serial0      0DE00800
*F5/0    192.67.67.53    Serial0      0DE00800
*F5/1    198.108.1.27    Ethernet0    AC0004006508AC00040002340800
*FE/0    198.108.13.28   Ethernet0    AC0004006508AC00040002340800
```

Figure 7.6 Viewing an example of the IP routing cache displayed from the use of the show ip cache command

When examining the IP routing cache illustrated in Figure 7.6 note that the asterisk (*) prefix designates valid routes, the destination field column contains the destination IP address, the Interface column indicates the interface by which a destination address is reached, while the column labeled 'MAC Header' contains the applicable MAC headers in hex notation. By examining the routes in the IP routing cache you can determine those routes that will be fast switched. Because fast switching is faster than the conventional router switching process, this will indicate those routes where the flow of packets towards the destination address have a minimum of initial latency when flowing outbound from your organization's router.

7.2.6 Displaying interface statistics

In the first section in this chapter we examined the use of the show interface command. Because the second section of this chapter is focused upon the display of IP-related information, we will now turn to the tailored version of the show interface command that provides such information. That command is the show ip interface command. The format of that command is:

show ip interface [<interface>]

Figure 7.7 illustrates an example of the use of the show ip interface command using the interface parameter of serial Ø.

When examining the resulting display shown in Figure 7.7, if you compare this illustration to the results from the use of the show interface command contained in Figure 7.1, you will note the former contains a few similar fields. However, a large number of fields are now IP oriented and can assist users in determining the potential ability of an existing configuration to support QoS.

```
Serial Ø is up, line protocol is up
  Internet address is 198.78.46.1, subnet mask is 255.255.255.0
  Broadcast address is 255.255.255.255
  Address determined by nonvolatile memory
  MTU is 1500 bytes
  Helper address is 205.131.1.255
  Outgoing access list is not set
  Proxy ARP is enabled
  Security level is default
  ICMP redirects are always sent
  ICMP unreachables are always sent
  ICMP mask replies are never sent
  IP fast switching is enabled
  Gateway Discovery is disabled
  IP accounting is enabled, system threshold is 512
  TCP/IP header compression is disabled
  Probe proxy name replies are disabled
```

Figure 7.7 The use of the show ip interface command using the interface parameter of serial Ø

This is because the use of the show ip interface command will inform you of the status of TCP/IP header compression and whether or not IP accounting is enabled, with the latter facilitating your ability to examining statistics concerning the flow of data.

Note that the first line of the resulting display informs you that both the port and line protocol is up. The second line in the display informs you of the IP address assigned to the interface, information that can also be obtained from the use of the show interface command. The next few lines inform you of the broadcast address and the MTU, again representing information you can obtain through the use of the show interface command. However, the next twelve lines in the display provide unique or near-unique information that warrants discussion.

Helper-address

The helper-address references a destination IP address for UDP broadcasts which provides a DHCP relay capability. That is, a DHCP relay agent enables a client and server to be located on separate subnets. In a Cisco environment, if a DHCP server cannot satisfy a DHCP request from its own database, it can forward the DHCP request to one or more secondary servers. Those servers are defined through the use of the 'helper-address' command, resulting in the helper-address entry in Figure 7.7 indicating the IP address of a DHCP server to which UDP broadcasts packets are sent.

Outbound access list

The line following the helper address indicates whether or not an outbound access list is set. This line can serve as a reminder to check any outbound access list, especially if it is used in conjunction with a queuing method to ensure that the access list entries correspond to your queuing intentions.

Proxy ARP and Security

The next two lines indicate whether Proxy ARP is enabled for the interface and if IP Security Options (IPSO) are set for the interface. The IP Security Option setting can indicate the use of TACACST or a similar Authorization, Authentication and Accounting (AAA) server.

ICMP information

The three lines following the security level indicate ICMP information. The first line in this troika specifies whether redirects will be sent on the interface. The second ICMP information line specifies whether unreachable messages will be transmitted on the interface. The third ICMP-related line specifies whether mask replies will be sent on the interface.

Fast switching and gateway

As its name implies, the line beginning with 'IP Fast Switching' specifies whether fast switching is enabled on the interface. The following line indicates whether the discovery process was enabled for the interface. On a serial interface fast switching is usually enabled, while the Gateway Discovery is typically disabled.

IP accounting

The third line from the bottom indicates whether IP accounting is enabled for the interface. This line also indicates the threshold in terms of maximum number of entries.

TCP/IP header compression

The next to last line in Figure 7.7 indicates whether TCP/IP header compression is enabled or disabled. As noted earlier in this book the use of TCP/IP header compression can significantly reduce the overhead associated with the transmission of digitized voice whose 20 ms segments are transported in relatively short packets.

Probe proxy name

The last line in the display of interface statistics concerns the probe proxy name. This field indicates whether the function is enabled or disabled.

Now that we have an appreciation for available interface statistics let us continue our tour of IP-related show commands with the show ip route command.

7.2.7 Displaying the routing table

You can display the contents of the IP routing table through the use of the show ip route command. The format of this command is shown below:

```
show ip route [<network>]
```

The use of the optional (network) parameter results in the display of the specified network in the routing table.

The top portion of Figure 7.8 illustrates the results obtained through the use of the show ip route command without specifying a network number.

Note the output commences by first displaying codes applicable for interpreting the first column in the remainder of the display after the gateway of last resort is displayed.

After the gateway of last resort is identified, the remainder of the display can consist of up to five fields. The first field indicates the manner by which the route was derived, with the codes at the top of the display of the routing table

```
A. The use of the show ip route command without the network number:

Codes: I - IGRP derived, R - RIP derived, H - HELLO derived
   C - connected, S - static, E - EGP derived, B - BGP derived
   * - candidate default route

Gateway of last resort is 144.122.6.7 to network 144.119.0.0

I* Net 198.145.0.0 [100/1020300] via 144.122.6.6, 30 sec, Ethernet0
I Net 192.68.151.0 [100/160550] via 144.122.6.6, 30 sec, Ethernet0
I Net 198.18.0.0 [100/8776] via 144.122.6.7, 58 sec, Ethernet0
                    via 144.122.6.6, 31 sec, Ethernet0
E Net 198.128.0.0 [140/4] via 144.122.6.64, 130 sec, Ethernet0
C Net 144.122.0.0 is subnetted (mask is 255.255.255.0), 54 subnets
I     144.122.144.0 [100/1310] via 144.122.6.7, 78 sec, Ethernet0
C     144.122.91.0 is directly connected, Ethernet1

B. The display result obtained by the use of the show ip route command and
   the optional network argument:

Routing entry for 144.122.1.0
 Known via 'igrp 109', distance 100, metric 1200
 Redistributing via igrp 109
 Last update from 144.122.6.7 on Ethernet0, 35 seconds ago
 Routing Descriptor Blocks:
 * 144.122.6.7, from 144.122.6.7, 35 seconds ago, via Ethernet0
   Route metric is 1200, traffic share count is 1
   Total delay is 2000 microseconds, minimum bandwidth is 10000 Kbit
   Reliability 255/255, minimum MTU 1500 bytes
   Loading 1/255, Hops 0
```

Figure 7.8 Using the show ip route command to display IP routing table entries

serving as the column entry. By examining the entry in the first field you can note if a route is derived via the use of a routing protocol or was statically assigned. Concerning the latter, as previously indicated in this book, the use of static routes can avoid the occurrence of routing table updates and the delay to traffic that occurs during the table update process.

The second field in the table portion of the display specifies a remote network or subnet to which a route exists. This field consists of a pair of numbers enclosed in brackets that are separated by a forward slash (/) character. The first number represents the administrative distance of the information source, while the second number represents the metric for the route to the network or subnet.

The fourth field specifies the number of seconds since the network or subnet was last heard. As noted earlier in this book, the time permitted prior to a route being purged depends upon the protocol used for the route. The fifth and final field indicates the interface through which the remote network can be reached.

In the lower portion of Figure 7.8 the results obtained from the use of the show ip route command with a network argument is shown. Note that displaying the routing table for a specific route results in detailed information about the route in the form of what is referred to as routing descriptor blocks information being displayed. Thus, in the lower portion of Figure 7.8 you will note that the routing descriptor blocks information metrics includes the total delay in microseconds, minimum bandwidth, reliability, minimum MTU, loading and hops to the route.

7.2.8 Displaying protocol traffic statistics

Continuing our tour of the use of different show ip commands, we can use the show ip traffic command to display IP protocol statistics. The use of this command can be an important tool in ascertaining probable causes of some types of QoS problems without having to revert to the use of a protocol analyzer. For example, consider Figure 7.9, which provides a sample output, obtained from the use of the show ip traffic command.

If you note a high level of broadcasts or ARPs, with the latter also representing a broadcast, this could result in random delays adversely effecting a voice over IP application. You can also examine the number of checksum errors for various protocols to obtain an indication of the overall quality of your communications. However, you need to realize that the ip traffic statistics displayed are cumulative for all ports and not for a specific interface. Thus, you would need to revert to the use of the show interface command to obtain specific information applicable to a specific router interface.

In examining the entries in Figure 7.9 a few words are warranted concerning several items. First, under the IP statistics heading as well as the EGP statistics heading you will note the term 'format error'. Here the term format error represents a major error occurring in a packet, such as a header length that is an odd number of bytes. An entry for a bad hop count actually references the situation where a packet is discarded because its time-to-live (TTL) field in the IP header was reduced to zero. It is important to note that the use of traceroute will result in a planned packet discard so unless you use a protocol analyzer to examine and decode packets it is possible for the bad hop count counter to provide dubious information.

Also under the IP statistics heading is a counter for encapsulation failed. An encapsulation failure normally indicates that the router did not receive a replay to an ARP request. Due to this, the router did not transmit a datagram.

Another entry under IP statistics that warrants discussion is the counter for no route. A value is added to the no route counter when the router discards a datagram it did not know how to route. What this means is that there was no entry in the routing table which would allow the router to intelligently forward the datagram, hence it was sent to the great bit bucket in the sky.

Concerning ARP and HP probe statistics, it should be noted that a proxy reply is counted when a router transmits an ARP or Probe Reply on behalf of another host. Both ARP and Probe statistics include the number of proxy requests that were received by the router as well as the number of responses that were sent.

```
IP statistics:
 Rcvd: 78 total, 78 local destination
      0 format errors, 0 checksum errors, 0 bad hop count
      0 unknown protocol, 0 not a gateway
      0 security failures, 0 bad options
 Frags: 0 reassembled, 0 timeouts, 0 too big
      0 fragmented, 0 couldn't fragment
 Bcast: 28 received, 42 sent
 Sent: 44 generated, 0 forwarded
      0 encapsulation failed, 0 no route
ICMP statistics:
 Rcvd: 0 checksum errors, 0 redirects, 0 unreachable, 0 echo
      0 echo reply, 0 mask requests, 0 mask replies, 0 quench
      0 parameter, 0 timestamp, 0 info request, 0 other
 Sent: 0 redirects, 2 unreachable, 1 echo, 1 echo reply
      0 mask requests, 0 mask replies, 0 quench, 0 timestamp
      0 info reply, 0 time exceeded, 0 parameter problem
UDP statistics:
 Rcvd: 46 total, 0 checksum errors, 45 no port
 Sent: 18 total, 0 forwarded broadcasts
TCP statistics:
 Rcvd: 0 total, 0 checksum errors, 0 no port
 Sent: 0 total
EGP statistics:
 Rcvd: 0 total, 0 format errors, 0 checksum errors, 0 no listener
 Sent: 0 total
IGRP statistics:
 Rcvd: 63 total, 0 checksum errors
 Sent: 26 total
HELLO statistics:
 Rcvd: 0 total, 0 checksum errors
 Sent: 0 total
ARP statistics:
 Rcvd: 18 requests, 16 replies, 0 reverse, 0 other
 Sent: 0 requests, 8 replies (0 proxy), 0 reverse
Probe statistics:
 Rcvd: 0 address requests, 0 address replies
      0 proxy name requests, 0 other
 Sent: 0 address requests, 0 address replies (0 proxy)
      0 proxy name replies
```

Figure 7.9   Using the show ip traffic command to display IP statistics

Now that we have an appreciation of the use of the show ip traffic command and the information it displays let us continue our tour of the use of show ip commands by focusing on how we can monitor the effect of TCP header compression.

7.2.9 Monitoring TCP header compression

Through the use of the show ip tcp header-compression command we can display statistics concerning the effect of enabling compression. Figure 7.10

```
TCP/IP header compression statistics:
Interface Serial0: (passive, compressing)
  Rcvd: 5020 total, 2891 compressed, 0 unknown type, 0 errors
       0 dropped, 1 buffer copies, 0 buffer failures
  Sent: 4584 total, 3634 compressed,
       116295 bytes saved, 681973 bytes sent
       1.16 efficiency improvement factor
  Connect: 16 slots, 1483 long searches, 2 misses, 99% hit ratio
       Five minute miss rate 0 misses/sec, 0 max misses/sec
```

Figure 7.10 An example of the resulting display obtained from the use of the show ip tcp header-compression command

illustrates an example of the display resulting from the use of the show ip tcp header-compression command.

When examining the display resulting from the use of the show ip tcp header-compression command, note that it informs us of the interface on which header compression was enabled, which in this example was the serial Ø interface. Concerning the counters shown in Figure 7.10, several deserve a bit of elaboration so let us discuss the meaning of the less obvious counters.

The buffer copies counter indicates the number of packets that needed to be copied into biggest buffers for decompression. When compression is very efficient the decompression of the header will require the use of the biggest buffer.

The counter for buffer failures indicates the number of packets dropped due to a lack of buffers. If this metric becomes large it indicates that an expansion of router memory would be in order.

The metrics for the total and compressed counters on the received and sent sides indicate the compression efficiency for inbound and outbound traffic. For the sent side for which your router performs header compression, additional statistics indicates the number of bytes saved through compression and the efficiency improvement factor. The latter term represents the improvement in line efficiency due to the enabling of TCP header compression. In Figure 7.10 the 1.16 efficiency improvement factor indicates that the line efficiency is increased by 16 percent due to enabling TCP header compression.

Under the connect heading the term 'slots' is used to reference the size of the cache. The counter for long searches indicates the number of times the compression software has to look 'hard' to encounter a match, while the counter for misses indicates the number of times a match could not be performed. If this counter has a relatively large value, it more than likely indicates that the number of allowable simultaneous compression connections was set to too low a value. If you reset the value upward you would then periodically issue additional show ip tcp header-compression commands to determine if the miss rate was related to the number of allowable simultaneous compression connections.

The hit ratio, which is expressed as a percentage, indicates the number of times per 100 that the software found a match and was able to compress the header. While this is nice to know, it's the efficiency improvement factor that should be of primary interest as that metric indicates the overall efficiency of header compression.

The last line displayed in response to issuing the show ip tcp header-compression command provides additional information concerning misses. The 5-minute miss rate displays the computed miss rate over the prior 5 minutes and provides a mechanism to examine miss rate trends. The last counter on the line provides an indication of the peak or maximum misses on a per second basis.

APPENDIX: TESTING TOOLS

The purpose of this appendix is to reacquaint readers with the use of two core testing tools to facilitate determining the practicality of obtaining a QoS capability via an intranet or over the Internet. Those two tools are the Ping and Traceroute applications built into TCP/IP protocol stacks.

PING

The most common use of Ping is to verify connections to a remote computer or computers. You can also use the Ping utility to test both the computer name and the IP address. If the IP address is verified, but the computer name is not, you may have a name resolution problem. In this case, be sure that the computer name you are querying is in either the local host's file or in the DNS database.

If you do not receive a reply from the device you're pinging, this indicates a network failure between the local and remote hosts. In actuality, that failure could be due to the improper cabling of the host you are using to the network or, if connected properly, the gateway that provides connectivity off the local area network could be inoperative. To test these theories you can consider several operations.

First, you can Ping the loopback address which is 127.0.0.1 to determine if your host's TCP/IP protocol stack is operating correctly. Next, you can Ping a host on your network to determine if your host can access other hosts on the network. Assuming both were accomplished, you could then Ping the gateway to make sure it is functioning. Next, you can Ping one or more hosts off the network to ensure you can reach other networks from your local area network.

Although Ping is built into Cisco's IOS, sometimes its easier to use your PC to Ping a distant device. The following example illustrates the use of Ping to query the status of the host at IP address 198.78.46.8.

```
C:\>ping 198.78.46.8

Pinging 198.78.46.8 with 32 bytes of data:

Reply from 198.78.46.8: bytes=32 time=20ms TTL=128
Reply from 198.78.46.8: bytes=32 time=20ms TTL=128
Reply from 198.78.46.8: bytes=32 time=20ms TTL=128
Reply from 198.78.46.8: bytes=32 time=25ms TTL=128
```

From a QoS standpoint the use of Ping provides the round trip delay to the distant host. While this information is certainly valuable, a few words of caution are in order concerning the use of Ping.

If you use a host name, such as *www.whitehouse.gov*, the first Ping will require some additional time as the host name to IP address must be resolved. Because subsequent Pings can use the learned and cached IP address, it is a good idea to use Ping multiple times and discard its first series of results. Similarly, because the round trip delay is based upon the activity of other hosts' requests and responses flowing through the network, it is a good idea to set up a batch file to periodically issue Pings and pipe the results onto a file. If you attempt this action, remember to use double greater than (≫) signs for piping as this adds to the file. Otherwise, each piping operation would overwrite the prior operation.

In a Cisco router environment you can also use a Ping utility program. An example of the use of Ping on the router named Macon is shown below:

```
Macon#ping
    Protocol [ip]:
    Target IP address: 198.78.46.8
    Repeat count [5]:
    Datagram size [100]:
    Timeout in seconds [2]:
    Extended commands [n]: y
    Source address or interface: 172.16.23.2
    Type of service [0]:
    Set DF bit in IP header? [no]:
    Validate reply data? [no]:
    Data pattern [0xABCD]:
    Loose, Strict, Record, Timestamp, Verbose [none]:
    Sweep range of sizes [n]:
    Type escape sequence to abort.
    Sending 5, 100-byte ICMP Echos to 162.108.21.8,
        timeout is 2 seconds:
    !!!!!
    Success rate is 100 percent (5/5), round-trip
        min/avg/max =36/97/132 ms
    Macon#
```

Ping command field descriptions

The table on page 180 lists the Ping command field descriptions. As shown above, these fields can be modified with the use of the Ping command.

TRACEROUTE

As previously noted, we can use the Ping utility program to verify connectivity between devices as well as to determine the round trip delay. While Ping is an important tool in our war upon determining connectivity and round trip delay, if there are intermediate routers between source and destination we cannot determine where any problems might lie if the results of the Ping are abnormal. In this situation we would turn to the use of the traceroute command, which for Microsoft Windows users, can be used by its abbreviation of tracert.

The traceroute command can be used to discover the routes packets take to a remote destination, as well as where routing breaks down. The host (to include router) executing the traceroute command sends out a sequence of User Datagram Protocol (UDP) datagrams, each with incrementing Time-To-Live (TTL) values, to an invalid port address at the remote host.

First, three datagrams are transmitted, each with a TTL field value set to 1. The TTL value of 1 causes the datagram to 'timeout' as soon as it hits the first router in the path. That router then responds with an ICMP 'time exceeded' message indicating that the datagram has expired. Because the host issuing the traceroute command set a timer, it discovers not only the first router on the path to the destination but, in addition, the round trip delay to that router.

Next, three more UDP messages are sent, each with the TTL value set to 2. This causes the second router in the path to the destination to return ICMP 'time exceeded' messages. In addition, the timer on the host now provides three more round-trip delay times, this time, no pun intended, to the second router along the path to the destination.

This process continues until the packets reach the destination and until the system originating the traceroute has received ICMP 'time exceeded' messages from every router in the path to the destination. Since these datagrams are trying to access an invalid port at the destination host, the host will respond with ICMP 'port unreachable' messages indicating an unreachable port. This event signals the traceroute program to finish.

The goal behind the traceroute command is to record the source of each ICMP 'time exceeded' message to provide a trace of the path each packet took to reach the destination. Because the round-trip delay is also computed, it becomes possible to note where delays within a network are occurring.

The following example illustrates the use of the traceroute command on the router named Macon.

```
Macon#traceroute
    Protocol [ip]:
    Target IP address: 192.78.46.8
    Source address: 192.16.22.2
    Numeric display [n]:
    Timeout in seconds [3]:
    Probe count [3]:
    Minimum Time to Live [1]:
    Maximum Time to Live [30]:
    Port Number [33434]:
    Loose, Strict, Record, Timestamp, Verbose [none]:
    Type escape sequence to abort.
    Tracing the route to 192.168.40.2
            1 192.21.20.3 [AS 3] 4 msec 1 msec 4 msec
            2 192.20.10.6 [AS 3] 5 msec 4 msec 8 msec
            3 192.16.30.2 [AS 3] 16 msec* 18 msec
    Macon#
```

The following table lists the traceroute command field descriptions used in a Cisco environment.

Field	Description
Protocol [ip]:	Prompts for a supported protocol. Enter appletalk, clns, ip, novell, apollo, vines, decnet, or xns. The default is; ip.
Target IP address:	Prompts you for the IP address or host name of the destination node you plan to ping. If you specified a supported protocol other than IP, you need to enter an appropriate address for that protocol. Default: none.
Repeat count [5]:	Indicates the number of ping packets that will be sent to the destination address. Default: 5.
Datagram size [100]:	Indicates the size of the ping packet (in bytes). Default: 100 bytes.
Timeout in seconds [2]:	Indicates the timeout interval. Default: 2 (seconds).
Extended commands [n]:	Specifies whether or not a series of additional commands appears.
Source address or interface:	Specifies the interface or IP addresses of the router to use as a source address for the probes. The router will normally pick the IP address of the outbound interface to use.

Type of service [0]:	Specify the Type of Service (ToS). The requested ToS will be placed in each probe but there is no guarantee that all routers will process the ToS.
Set DF bit in IP header? [no]:	Specify whether or not the Don't Fragment (DF) bit is set on the ping packet.
Validate reply data? [no]:	Specify whether or not to validate the reply data.
Data pattern [0xABCD]	Specify the data pattern. Different data patterns are used to troubleshoot framing errors and clocking problems on serial lines.
Loose, Strict, Record, Timestamp, Verbose[none]:	IP header options. You can specify any combination. The traceroute command issues prompts for the required fields. Note that traceroute command will place the requested options in each probe. However, there is no guarantee that all routers (or end nodes) will process the options.
Sweep range of sizes [n]:	Each exclamation point (!) indicates receipt of a reply. A period (.) indicates the network server timed out while waiting for a reply. Other characters may appear in the ping output display, depending on the protocol type.

Other common characters and their meanings include:

Character	Meaning
U	Destination Unreachable error PDU received
N	Network Unreachable
P	Protocol Unreachable
Q	Source Quelch
M	Could not fragment
?	Unknown packet type

Success rate is xxx percent	Percentage of packets successfully echoed back to the router.
round-trip min /avg/ max = 1/2/4 ms	Indicates the round-trip travel time intervals for the protocol echo packets, including minimum/average/maximum (in milliseconds).

Field	Description
Protocol [ip]:	Prompts you for a supported protocol. Enter appletalk, clns, ip, novell, apollo, vines, decnet, or xns. The default is: ip.
Target IP address:	Enter a host name or IP address. There is no default.
Source address:	The interface or IP addresses of the router to use as a source address for the probes. The router will normally select the IP address of the outbound interface to use.
Numeric display [n]:	Defines the use of a symbolic or numeric display. The default is to have both a symbolic and numeric display; however, you can suppress the symbolic display.
Timeout in seconds [3]:	The number of seconds to wait for a response to a probe packet. The default is 3 seconds.
Probe count [3]:	The number of probes to be sent at each time to live (TTL) level. The default count is 3.
Minimum Time to Live [1]:	The TTL value for the first probes. The default is 1, but it can be set to a higher value to suppress the display of known hops.
Maximum Time to Live [30]:	The largest TTL value that can be used. The default is 30. The operation of the traceroute command terminates when the destination is reached or when this value is reached.
Port Number [33434]:	The destination port used by the UDP probe messages. The default Port is 33434.
Loose, Strict, Record, Timestamp, Verbose [none]:	IP header options. You can specify any combination, such as loose and timestamp. The traceroute command issues prompts for the required fields. Note that traceroute command will place the requested options in each probe; however, there is no guarantee that all routers (or end nodes) will process the options.

INDEX

Printed and bound in the UK by
CPI Antony Rowe, Eastbourne

Printed and bound by CPI Group (UK) Ltd, Croydon, CR0 4YY

27/10/2024

14580218-0001